A Bibliography of Telecommunications and Socio-Economic Development

A Bibliography of Telecommunications and Socio-Economic Development

Heather E. Hudson

Artech House

Library of Congress Cataloging-in-Publication Data

Hudson, Heather E.
A bibliography of telecommunications and socio-economic development / Heather E. Hudson.
p. cm.
Includes index.
ISBN 0-89006-288-9 (pbk.)
1. Telecommunication--Bibliography. 2. Telecommunication policy--Bibliography. 3. Economic development--Bibliography. I. Title.
Z7164.T28H83 1988
[HE7631]
016.384--dc19 87-35089
CIP

Published by
ARTECH HOUSE, INC.
685 Canton Street
Norwood, MA 02062

International Standard Book Number: 0-89006-288-9
Library of Congress Catalog Card Number: 87-35089

10 9 8 7 6 5 4 3 2 1

Contents

Acknowledgements

Many people contributed to the compilation of this bibliography. Research assistants included Janie Briesemeister, Henry Geddes, Linda Krueger, Louis Leung, Judith Swain, and Lynn York. The searches of computerized data bases were done at the Perry Castenada Library of the University of Texas at Austin with the assistance of Ronald Seeliger, communications librarian. The initial data base was compiled on the University of Texas computer system. Facilities for downloading and final editing of the data base on microcomputers were provided by the Center for the Study of Communication Technology and Society in the College of Communication, directed by Frederick Williams. Partial funding for the project was provided through grants from the University Research Institute and the College of Communication of the University of Texas at Austin.

Introduction

1. OVERVIEW

This book is designed for people who need to find information on the relationship of telecommunications to economic and social development. This field has attracted growing interest in the 1980s, as innovations in technology, a proliferation of new services, and changes in the telecommunications industry have coincided with a greater awareness of the contribution of telecommunication and information technologies to development in the broadest sense, including economic growth and improved access to education and social services. Since the 1970s, the constituents for this information now include officials in telecommunications administrations and ministries of communication, senior officials in national planning ministries, multilateral and bilateral agencies that fund telecommunications projects and applications, and corporate strategists in the telecommunication industry who seek to understand and to enter new markets, as well as researchers from several disciplines. The "invisible college" in this field has expanded to include not only engineers and economists, but also researchers with backgrounds in finance, social psychology, anthropology, regional planning, education, and rural sociology.

While heightened visibility has greatly enriched the field, it has also made the search for relevant material both more complex and more rewarding. There are certainly many more studies now than there were five years ago, but they are still scattered in academic journals, professional publications, trade press, and conference proceedings of numerous disciplines. Researchers may not uncover the most relevant studies simply by reviewing the literature in their field. However, even consulting a data base may not be sufficient. References may be scattered throughout so many fields that it is necessary to search numerous data bases that cover social sciences, engineering, education, and international development to find relevant material.

This bibliography compiles citations from numerous data bases and other sources to assist researchers and planners concerned with the role of telecommunications in socio-economic development.

2. CONTENTS AND SOURCES

The data base includes materials in the field through 1985. Entries published in 1986 are included, but 1986 material may not be as comprehensive due to the time lapse between publication and entry into some data bases, and between compilation of this bibliography and publication. Data bases searched included the following:

ABI/Inform (Data Courier)
Books in Print Index (R.R. Bowker)
Dissertation Abstracts Online (University Microfilms International)
Economic Literature Index (American Economic Association)
ERIC (Educational Resources Information Center)
Foreign Trade and Economic Abstracts (Netherlands Foreign Trade Agency)
GPO Monthly Catalog (US Government Printing Office)
InfoTrac (Information Access Company)
INSPEC (Institute of Electrical Engineers, UK)
NTIS (National Technical Information Service, US Department of Commerce)
PAIS International (Public Affairs Information Service)
Social Scisearch (Institute for Scientific Information)
Sociological Abstracts (Sociological Abstracts, Inc.)
Soviet Science and Technology (IFI/Plenum Data Corporation)

In addition, the author's earlier bibliography on telecommunications and development (published in Hudson, 1979, and updated through 1982 in *When Telephones Reach the Village* (Hudson, 1984)) was included, with additional materials from the author's extensive collection of studies, monographs, and conference proceedings.

A set of search terms was compiled to extract citations that address the relationship of telecommunications to socio-economic development. Telecommunications was limited to interactive point-to-point services (e.g., telephone, telex, new technologies and services such as videotext and electronic mail). Citations that concerned only mass media were excluded, as they are well represented in other literature. Technologies that deliver interactive services such as satellites, microwave, and two-way radio were included. Particular effort was made to find sources of information on rural telecommunications applications and effects. In addition to developing country-based rural studies, materials on rural projects and research in industrialized countries were included. Studies of primarily urban effects in industrialized countries were excluded, as they are also well represented in existing sources and are not as likely to be relevant to developing country environments.

The bibliography contains 1124 entries, each with the following information:

author
title
publisher information or periodical citation
key words

The set of key words used for classifying entries includes the following:

Geographical Regions

Africa
Caribbean
Europe
Latin America
North America
Pacific
World
(Individual countries are also cited.)

Technologies and Media

audiovisual
broadcast
cable
computer
digital
fiber
film
information
mass media
microwave
radio
satellite(s)
technology
telecommunication(s)
teleconferencing
telematics
telephone
teletext
television
telex
two-way radio
video
videotext

Content and Applications

agriculture
business
cost
cultural
data
deregulation
economic(s)
education
evaluation

finance
health
history
international
investment
legal
management
model
orbit
planning
policy
political
regulation
remote
research
rural
social
spectrum
statistics
technical
theory
training
transportation
urban

Organizations

AID (US Agency for International Development)
ECOWAS (Economic Organization of West African States)
INTELSAT
NASA (National Aeronautical and Space Administration)
ITU (International Telecommunication Union)
PANAFTEL (Pan African Telecommunications Network)
UNESCO
United Nations
WARC (World Administrative Radio Conference)
World Bank

For journal articles, the last line of the citation (preceding the key words) generally contains the volume, number (N), and pages (P) of the article. Some citations contain additional information from the data base source. Those from Dissertation Abstracts and Masters Abstracts list document reference numbers. Citations from the ERIC data base contain document numbers beginning with ED (educational document) or EJ (educational journal). GPO documents list a GPO reference number. Citations from the NTIS data base include NTIS document numbers beginning with A, N, or P.

Key word indices are provided to assist the reader in finding entries on specific topics. Each index lists by number all the entries with that particular key word. Cross-indexing may be accomplished by finding the set of entries which contains both key words. For example, a reader looking for articles on satellites in Africa can look up SATELLITES and AFRICA, and note all entries listed in both indices.

Pilot testing of the bibliography indicated that a single listing with multiple key words offered the greatest flexibility and ease of access for the reader. Separate listings under geographical region, technology, and service, for example, would have resulted in hundreds of duplications and cross-references.

The bibliography was initially compiled on a mainframe computer. However, it was transferred to a microcomputer and then reformatted and edited by using dBase III. An on-line version of the bibliography enables users to search the data base by using any combination of key words and other parameters (author, title, date, *et cetera*). The floppy disk version is available from the publisher, and requires use of dBase III software on an IBM XT, AT, or compatible computer.

Further Updates

An advantage of the computerized format is that the bibliography may be easily updated and expanded. Updates will be issued separately to avoid the cost of republishing the entire data base. Floppy disk versions may also be available.

While searches of data bases yield a wealth of material, there are many other sources of information that are more difficult to find. Conference papers, foreign language studies, and materials produced by government and international agencies tend not to be included in most data bases. Readers are invited to submit entries for the updates. Entries should include the information provided in the citations in this bibliography. A copy of the document or an annotation of its contents would be very helpful. Please send citations and other comments to the author in care of Artech House.

3. RECENT DEVELOPMENTS

Several factors have resulted in increased attention to telecommunications during the 1980s. At the beginning of the decade, the International Telecommunication Union (ITU) sponsored a series of studies to attempt to quantify the socioeconomic benefits of investment in telecommunications. World Communications Year in 1983 focused global attention on the importance of telecommunications in the development process. In that year, the ITU published a synthesis of its studies, titled *Telecommunications and Development* (Pierce and Jequier, 1983). Several sessions at the ITU's Forum 83 in Geneva were devoted to telecommunications and development. The ITU also held regional seminars on this theme for developing country planners and policy makers in Costa Rica, Togo, and Malaysia.

During the same year, the World Bank also published a comprehensive overview of the field from an economic perspective, titled *Telecommunications and Economic Development.* The year 1983 also saw the establishment of the International Commission on Worldwide Telecommunications Development, known as the Maitland Commission, for its chairman, Sir Donald Maitland. The Maitland Commission was mandated by a resolution from the ITU's 1982 Plenipotentiary Conference in Nairobi to "recommend ways to stimulate the expansion of telecommunications across the world."

The Commission's report, *The Missing Link,* published in 1985, pointed out the extent of underinvestment in telecommunications in the developing world, and suggested several strategies for both increasing investment and obtaining greater benefits from telecommunications utilization.

The 1982 Plenipotentiary Conference also mandated the creation of a small Telecommunications Economics Division within the Technical Cooperation Department of the ITU. This unit issued a second set of studies in 1986, titled *Information, Telecommunications and Development.*

Another factor stimulating interest in the field has been the rapidity of technological and institutional change during the 1980s. Several developing countries joined the satellite family, with the launch of India's domestic satellite in 1983, followed by Brazil, Mexico, and the Arab nations in 1985. Indonesia is now in its second generation of with PALAPA B, a regional system. More than 20 developing nations now lease capacity from INTELSAT for domestic communications.

The convergence of telecommunications and computing technologies has resulted in a variety of new services, ranging from micro earth stations capable of transmitting and receiving low volumes of data very cheaply to integrated services digital networks (ISDN) that will be electronic superhighways. As the variety of equipment and the computing power available through miniaturization increase, planners and researchers are asking how or even whether these innovations can contribute to the development process.

Changes in the structure of the telecommunication industry are also having global effects. The deregulation of US telecommunications has raised many questions about the benefits to small users and rural customers, e.g., whether there will be sufficient incentive to keep rates affordable and service adequate for rural and other less attractive customers. We face a paradox of technological change that now makes it possible to communicate with virtually anyone anywhere, and institutional change that may make these services less widely available or prohibitively expensive. Lack of adequate telecommunications infrastructure is a barrier to decentralization of economic activities and services, and may create yet another incentive for migration to overburdened cities.

Industrialized countries are exploring new methods for operating their telecommunications systems. There is a trend toward privatization of government-owned or operated telecommunications facilities, and, in some cases, also liberalization, i.e., the introduction of competition. Developing countries are being encouraged to operate autonomous government telecommunications administrations so that there is more incentive for efficiency, and earnings may be retained to improve and expand the networks, rather than being used to subsidize other services such as the post, or simply generating additional revenue for the national treasury.

International competition is another issue that creates both opportunities and problems in extension of rural telecommunications. New international satellite systems competing with INTELSAT may drain revenue from the INTELSAT system, resulting in higher average costs for using the network. On the other hand, new international services may offer developing countries attractive options in lower cost or more flexible domestic and regional services. However, INTELSAT today is a much more flexible entity than it was in the early 1980s, and many of its new offerings are aimed at providing new services and lowering costs for developing countries.

Another trend is the diffusion of new technologies that blur traditional distinctions between telecommunications, broadcasting, and computing, and offer alternative means of obtaining and disseminating information. A microcomputer with modem in a developing country can access data bases in industrialized countries. A satellite earth station can receive valuable news and weather information, and may also pick up broadcasting signals from neighboring countries. A video cassette recorder (VCR) also brings foreign information into countries, regardless of their national policies. The diffusion of VCRs and satellite antennas offers dramatic evidence that video is no longer strictly a broadcast service, and that governments basing their communications policies on traditional definitions of telecommunications and broadcasting may be missing the most important ways in which their citizens obtain information.

4. A VARIETY OF PERSPECTIVES

A variety of disciplines, each operating quite independently, contributed to early knowledge about the role of telecommunications in socio-economic development. Although the number of studies in this field has continued to grow, there is not yet a convergence of theory and methodology, nor even widespread recognition that researchers are contributing to a common pool of expertise. We must still search data bases covering engineering, economics, political science, sociology, education, anthropology, and agriculture—as well as communications—to find studies related to the field.

There are many ways to classify research in this field. Theoretical studies may build models to show the relationship of telecommunications investment to economic development, for example, or propose methodologies to quantify benefits or isolate relevant variables. Applied research includes case studies, sector studies, and national studies that collect data or analyze existing data to attempt to discover actual relationships between telecommunications investment or use and measurable indicators of development.

In addition to crossing several disciplines, research on telecommunications and development may approach the topic from several perspectives. The major approaches include:

The applications approach. Studies examining the contribution of telecommunications in a particular application, such as distance education, health care delivery (telemedicine), agriculture, and corporate management (electronic mail, teleconferencing).

The economic approach. Studies examining economic results of telecommunications investments, usually on a project or case study basis, e.g., contributions of telecommunications to increased productivity, inventory management, transportation substitution, cost-benefit and cost-effectiveness studies, *et cetera.*

The sociological approach. Studies of the effects of new technologies in rural areas growing out of the rural sociology tradition; studies on information workers and on technology in the work place.

The policy approach. Examination of issues such as access to technology, changes in industry structure and pricing, planning of institutional or national telecommunications systems.

These studies may also be classified according to whether they are carried out in a developing country or industrialized country context.

Developing countries.

Rural areas —often home to majority of the population; generally agricultural (cash crops or subsistence farming);

Remote regions—isolated by mountains, water, desert, *et cetera*; may include mining, petroleum, fishing, tourism applications, as well as provision of basic services.

Industrialized countries.

Rural regions (e.g., agricultural areas);

Remote regions (e.g., Alaska, northern Canada, the Australian outback).

This bibliography contains studies using each of these approaches and from both industrialized and developing country contexts. The major types of studies represented are given below.

Case Studies of Various Sectors

One of the most prevalent types of research is the case study of the role of telecommunications in a particular sector, such as agriculture, education, or industry. Early studies examined agriculture in the South Pacific, businesses in Kenya, and distance education in Alaska. More recent studies report research from Peru, the Philippines, Senegal, and several other developing countries.

These studies are particularly valuable because they give concrete evidence at the micro level of the role of information technology and telecommunications, which helps us to understand the relevance (and limitations) of current theory. However, the generalized applicability of these studies is limited because of the variations in methodology used, and in the organization and structure of the sectors in different countries. A theme that does recur is that these latter factors may be the most important in predicting telecommunications benefits. A well managed operation with trained personnel, access to credit, and a choice of markets or suppliers is likely to gain more from the use of telecommunications than a subsistence operation or activity with poorly trained and managed staff and limited access to resources. These structural factors are analyzed in more detail under planning and policy studies below.

Studies on Rural North America

Another important source of theory and evidence is studies on the effect of new information and telecommunications technologies in rural North America. There are obvious differences in scale, wealth, and degree of mechanization between agriculture in North America and the developing world, but the assessments of results and the methods of research appear highly relevant for developing country studies. Interestingly, there seems to have been very little contact beween the rural sociologists and agricultural economists in North America, who are shifting their attention to information technologies, and researchers who have been studying the effect of telecommunications in developing countries.

Analysis of the Information Sector

The primary information sector includes those parts of the economy that are directly involved with the production, processing, and distribution of information, and the production and maintenance of information technologies. The secondary information sector includes those aspects of other sectors that are primarily involved with information activities. Thus, education, publishing, banking, and computer assembly would be part of the primary information sector, whereas the accounting and medical records activities of health care delivery, the inventory and marketing aspects of automobile production, and the analysis of drill site data in the petroleum industry would be part of the secondary information sector.

The first studies of the information sector analyzed the economies of OECD countries. Recent studies have examined the primary information sectors of some developing countries, primarily newly industrialized countries. Few attempts have been made to extricate the components of the secondary information sector, however. This is a difficult and complex task because it involves an assessment of how much activity in all the other components of the economy—i.e., agriculture, forestry, and mining; manufacturing; and noninformation services—is actually information-related. Yet, in the middle-income and poorer developing countries, these sectors are likely to be by far the most important in terms of output and employment.

Telecommunications System Planning and Needs Assessment

Telecommunications system planning for developing countries generally includes a projection of the expected direct benefits (revenues) and indirect benefits of the investment. There is growing interest in assessing the rural benefits because most developing countries have a high percentage of their population and economic activity in the countryside.

Policy Studies

Studies of telecommunications policies that relate to developing countries are another source of insights on the role of telecommunications in the development process. These studies tend to focus on:

- the role of information in national policies, negotiations, and trade;
- analysis of developing countries' requirements, positions, and participation in international telecommunications policy fora such as the ITU, UNESCO, and GATT;
- analysis of the telecommunications context: the social, political, economic, and cultural factors that may influence how telecommunications systems are used and with what benefit in developing countries.

Studies on the roles of developing countries in telecommunications policy formulation in the mid-1980s have focused primarily on international satellite policy. Two major issues have been the negotiations over access to the geostationary orbit (Space WARC) and international competition with INTELSAT.

Project Studies

A final source of research findings on the role of telecommunications in the development process is evaluation of pilot projects and other field studies. Many of these are funded by multilateral and bilateral aid agencies. The World Bank and the US Agency for International Development (AID), for example, require project evaluations that can contribute to better understanding of the benefits of telecommunications in specific sectors such as health, education, or agriculture. Recent satellite-based development projects were sponsored by AID and INTELSAT. The AID Rural Satellite Program supported projects linking agricultural universities in Indonesia, extending distance education facilities of the University of the West Indies, and providing audio teleconferencing for health services as well as public telephone service in rural Peru. INTELSAT's Project Share included distance education projects in China and the South Pacific, as well as several shorter term experiments and demonstrations.

5. INTEGRATING TELECOMMUNICATIONS PLANNING WITH DEVELOPMENT PLANNING

The Maitland Commission concluded: "It is our considered view that henceforward no development programme of any country should be regarded as balanced, properly integrated, or likely to be effective unless it includes a full and appropriate role for telecommunications."

Yet, even if development planners concur with the Maitland Commission's recommendations, they need substantially more guidance on how to incorporate telecommunications into their programs. Among the issues that need to be addressed are:

Trade-offs between dedicated systems and multipurpose systems. Whether to install a dedicated two-way radio system for medical services or a new railroad, for example, or to extend the national public network to provide these services.

Trade-offs between improving the quality of the existing network and extending the network to new subscribers. In many countries, inadequate maintenance and excess demand result in poor quality of service in the existing network. Adding new lines or extending to new communities will simply compound the problem. However, waiting until the network is fully upgraded may result in delays that keep the country from benefiting from telecommunications services. How can a planner determine what mix of upgrading and extension of services is appropriate? Under what circumstances would bypass or dedicated networks be appropriate?

Development-based planning criteria. Telecommunications planners tend to use population and administrative divisions as major criteria in planning extension of service. For example, the first priority may be to link all the provincial capitals. A second priority may be to add regional headquarters, district headquarters, *et cetera*, down to the village level. While this approach can be the best starting point, it may need modification to respond to priority needs. For example, a new project such as a mine or mill may need telecommunications, even if it is located far away from a town (or possibly because it is so isolated). Population centers, no matter how small, that are on major transportation routes, such as highway intersections, river landings, or border crossings, may require priority service. Community of interest needs to be taken into account in planning the network so that traffic can be easily routed between centers, even if they happen to be in different districts or provinces. Finally, telecommunications planning needs to be integrated with national and regional development planning so that communications services can be provided to support activities such as the building of new towns, opening of new tourist areas, or extension of health and education services.

Flexibility. These approaches require flexibility in planning and in technology. They also require coordination between the telecommunications authority and government planners so that network plans and procurements can be based on development priorities. Technical flexibility may also be required—for example, using small satellite earth stations to provide "instant service" to remote installations, using subscriber radio links to reach outlying communities, and adding drops from trunk networks to serve subscribers along the route.

Thus, flexible planning based on coordination with other government ministries and with representatives at regional and subregional levels is needed to ensure that telecommunications investments achieve maximum developmental results.

6. A RESEARCH AGENDA

The above discussion of questions facing planners suggests topics for additional research. In addition, there is a need to attempt to develop some standardization in methodologies so that case studies may be replicated in different countries and for different sectors, and cost-benefit studies may be based on a generally accepted technique for estimating telecommunications costs and quantifying benefits.

Such standardization is not meant to suppress innovation. There is a need to look for new methodologies (or methodologies from other disciplines) that may shed more light on the role of telecommunications. For example, the concept of "value added" that is now being applied in office automation studies appears promising for telecommunications studies. Also, modeling techniques adapted from applied economics may be useful to estimate the effect of telecommunications investments.

Primary information sector studies have now been undertaken for several developing countries. However, there is also a need to analyze the extent of the secondary information sector in developing economies, i.e., the contribution of information activities in other sectors such as agriculture, fisheries, resource development, manufacturing, *et cetera*. More theoretical work also is needed to increase our understanding of the role of information and, by extension, telecommunications in economic development. In particular, how does information access and utilization contribute to the productivity of resource industries and manufacturing, which are the mainstays of most developing economies?

Finally, as noted above, the diffusion of "new technologies" raises many questions for researchers and planners. How can new technologies be harnessed for development? How can countries that have invested in satellite communications, for example, obtain the maximum developmental benefits? Can microcomputers, cellular radios, micro earth stations, to name a few, return benefits if adopted early? Can new strategies in management, rate structures, and flexible policies of equipment ownership and special services accelerate diffusion and productive use of new technologies?

7. BRIDGING THE GAP BETWEEN POLICY AND RESEARCH

The research agenda given above is intended to build on research to date in order to enrich our understanding of the role of telecommunications in the development process. However, this knowledge will be much more valuable if it is shared with telecommunications planners and policy makers. Researchers must communicate not only across disciplines, but across professions, to reach those who need their findings. This book is intended for researchers and planners alike, with the goal of bridging the gap between them.

A Bibliography of Telecommunications and Socio-Economic Development

1 ABEL, E.
GLOBAL INFORMATION: THE NEW BATTLEGROUND
1982
POLITICAL COMMUNICATION AND PERSUASION
1 N4 347-57

INTERNAT INFO POLICY SOC
POLIT

2 ABLER, R.
THE TELEPHONE AND THE EVOLUTION OF THE AMERICAN METROPOLITAN SYSTEM
IN I. DE SOLA POOL (ED) THE SOC IMPACT OF THE TELEPHONE
1977
MIT PRESS: CAMBRIDGE, MA.

TELEPHONE SOC URBAN HIST
N AMER

3 ABRAMSON, JOAN
SATELLITES FOR RURAL DEVELOPMENT: AN APPROPRIATE TECHNOLOGY
1983
PAC ISLAND COMM J
12 N2 P15-28

PAC SAT TELECOMM RURAL

4 ABRAMSON, NORMAN
ASTIFONET WITHIN THE FRAMEWORK OF THE REGIONAL NETWORK FOR THE
EXCHANGE OF INFORMATION AND EXPERIENCE IN SCIENCE AND TECHNOLOGY
1984
UNESCO: PARIS

ASIA PAC INFO DATA

5 ABRAMSON, NORMAN
PALAPA FOR DATA COMMUNICATION
FINAL REPORT/UNDP PROJECT
1979, AUGUST
P1-26
FINANCIAL INFORMATION SYSTEMS

DATA FIN

6 ABRAMSON, NORMAN
SATELLITE DATA NETWORKS FOR NATIONAL DEVELOPMENT
1984, MARCH
TELECOMM POLICY (GB)
8 N1 P15-28

DATA ECON SAT TELEPHONE
BROADCAST DIGITAL

7 ACADEMY FOR EDUCATIONAL DEVELOPMENT, INC.
EDUCATIONAL TECHNOLOGY AND THE DEVELOPING COUNTRIES:
A HANDBOOK
1972
ACADEMY FOR EDUCATIONAL DEVELOPMENT: WASHINGTON, D.C.

EDUC INTERNAT MEDIA

8 ACADEMY FOR EDUCATIONAL DEVELOPMENT, INC.
SATELLITE OPTIONS FOR RURAL TELECOMM IN DEVELOPING COUNTRIES
1981, JUNE
PB81-244766
AED: WASHINGTON, D.C.

SAT TELECOMM RURAL PLAN

9 AEU
SOLVING THE PROBLEMS OF RURAL TELECOMMUNICATION
1980, MAY
AEU (JAPAN)

RURAL TELECOMM ASIA JAPAN

10 AGENCY FOR INTERNATIONAL DEVELOPMENT
PROJECT PROFILES. AID STUDIES IN EDUCATIONAL TECHNOLOGY
AND DEVELOPMENT COMMUNICATION
UNDATED
ACADEMY FOR EDUCATIONAL DEVELOPMENT: WASHINGTON, D.C.

RURAL SAT MEDIA TV
RADIO HEALTH

11 AGGAD, LEYLA
TOWARDS AN ARAB OPEN UNIVERSITY THROUGH THE
ARAB SPACE NETWORK
1980
ALEPPO UNIV PRESS: SYRIA

ARAB M EAST EDUC SAT
SYRIA

12 AGRAWAL, BINOD C.
ANTHROPOLOGISTS CONTRIBUTE TO INDIA'S SITE EXPERIMENT
1982, MARCH
INTERMEDIA
10 N3 P32-36

INDIA SAT RES EVAL
INFO ECON POLIT

13 AGRAWAL, BINOD C.
SATELLITE INSTRUCTIONAL TV EXPERIMENT: TV COMES TO VILLAGE--
AN EVALUATION OF SITE
1978, OCTOBER
INDIAN SPACE RES ORGANIZATION: BANGALORE, INDIA

INDIA SAT TV EDUC
EVAL

14 AHMED, K., DESSOUDI, A. AND POOL, I.D.S.
COMMUNICATION SYSTEM IN RURAL EGYPT
REPORT #11, COMMUNICATION NEEDS FOR RURAL DEVELOP. RESEARCH PROJECT
1980
CAIRO UNIVERSITY AND MIT

RURAL EGYPT

15 AIYA, S.V.C.
TELECOMM SERVICES FOR RURAL INDIA
1982, DEC
J. INST. ELECTRON. & TELECOMM ENG. (INDIA)
28 N12 P738-52

INDIA RURAL TELECOMM

16 AIYA, S.V.C.
TELECOMM SYSTEMS FOR INDIAN DEVELOPMENT
ADMINISTRATION
1985 APRIL
TELECOMM
52, N4, P240-6

INDIA TELECOMM SOC ECON

17 AIYA, S.V.C.
THE RURAL TELECOMMUNICATION PROBLEM
1980, DEC.
INDIAN AND EAST. ENG.
122 N12

RURAL TELECOMM INDIA

18 AIYA, S.V.C.
THE RURAL TELECOMMUNICATION PROBLEM IN
DEVELOPING COUNTRIES
1980, FEB.
TELECOMM JOURNAL (ENGLISH ED.) (GENEVA)
47 N2

RURAL TELECOMM

19 AJAYI-DOPEMU, Y.
VISUAL AIDS AND THE ENHANCEMENT OF COMMUNICATION IN AFRICA
1982, OCT.
J OF EDUCAL TV
8, N3 P203-209

AF AV CULT
SOC TV

20 ALBERTSON, LESLY, A.
TELECOMMUNICATIONS AS A TRAVEL SUBSTITUTE: SOME PSYCHOLOGICAL,
ORGANIZATIONAL AND SOCIAL ASPECTS
1977
JOURNAL OF COMMUNICATION
27 N2 P32-43

TELECOMM TELECONF SOC

21 ALEGRETT, JOSE L.
INTELSAT FACING NEW CHALLENGES IN THE GLOBAL INFORMATION ECONOMY
1985, DEC 19
SPEECH/SENIOR SEMINAR, FOREIGN SERVICE INSTITUTE, U.S. DEPT OF STATE
INTELSAT: WASHINGTON, D.C.

WORLD INTELSAT INFO ECON
FIBER CABLE

22 ALFIAN, AND CHU, GODWIN C.
SATELLITE TELEVISION IN INDONESIA
1981
EAST-WEST CENTER: HONOLULU, HAWAII

INDONESIA SAT TV

23 ALFIAN, ET AL
MEDIA AND DEVELOPMENT IN INDONESIA
ED235775
ERIC DOC REPRO SERV

INDONESIA MEDIA RURAL AGRIC

24 AMERICAN TELEPHONE AND TELEGRAPH CO.
THE WORLD'S TELEPHONES
ANNUAL
AT&T: BEDMINSTER, NJ

WORLD STAT TELEPHONE

25 ANASIUDU, RAFAEL CUKWUBANA
THE BENEFITS AND PROBLEMS OF AFRICAN COUNTRIES' PARTICIPATION IN INTELSAT
1983
UNIVERSITY MICROFILMS INTERNATIONAL: ANN ARBOR, MICHIGAN

AF INTELSAT

26 ANDERSON, C. AND RAPPAPORT, W.
TECHNOLOGY APPLICATIONS IN RURAL HEALTH CARE SYSTEMS
1977, MAY
MITRE CORPORATION; MCLEAN, VA

US VERMONT TECH HEALTH
REMOTE TELEPHONE

27 ANDERSON, JOHN W.
THE ECONOMIC, LEGAL AND SCIENTIFIC IMPLICATIONS OF DBS
1985, FEB
COMM & THE LAW
7 N1 P3-25

ECON LEGAL SAT TECH

28 ANDERSON, R.E.
SATELLITES MEET RURAL NEEDS
1984, MARCH
TEL ENG & MGT
88 N5 P72-5

SAT RURAL

29 ANE, C.K.
TECHNOLOGY FOR TRANSMISSION MEDIA IN RURAL TELECOMMUNICATION
1979, DEC
TELECOMM (INDIA)
29 N2

TECH INDIA RURAL TELECOMM
SAT

30 APETE, KOUASSI
LE RESEAU DE TELECOMMUNICATION PAN AFRICAIN PANAFTEL:
PROMESSES ET REALITES
1983
PAPER (IN FRENCH)/SYMPOS ON POLICY, FIN AND ECON ASPECTS
OF TELECOMM
INTERNATIONAL TELECOMMUNICATION UNION: GENEVA

AF PANAFTEL TELECOMM POLICY
FIN ECON

31 APPEL, GARY L. AND SCHLENKER, ROBERT E.
LONG-RANGE SATELLITES -- A VIABLE WAY FOR GROUP PRACTICES
TO PROVIDE REMOTE RURAL CARE
1978 JAN-FEB
MEDICAL GROUP MGT
25 N1 P56-58

SAT RURAL HEALTH REMOTE

32 APTED, MAURICE
PACIFIC OUTREACH
UNIVERSITY OF THE SOUTH PACIFIC, SATELLITE COMMUNICATIONS PROJECT
1980, JANUARY

SAT PAC

33 APTED, MAURICE
THE UNIVERSITY OF THE SOUTH PACIFIC'S EXPERIENCE: DISTANCE
TEACHING USING THE UNIVERSITY'S RADIO SATELLITE NETWORK
1980, OCTOBER
P1-19
THE UNIVERSITY OF THE SOUTH PACIFIC

SOUTH PAC SAT RADIO EDUC
ARGIC HEALTH MGT

34 ARELLANO, M.H.
CONSIDERACIONES SOCIALES EN LA PLANIFICACION DEL DESARROLLO
DE LAS TELECOMM EN LAS REGIONES DISTANTES Y DESFAVOR...
1978
WORKSHOP ON TELECOMM DEVELOPMENT IN ISOLATED & UNDERPRIVILEGED AREAS

L AMER TELECOMM SOC RURAL

35 ARGUMENDO, ALCIRA
THE NEW WORLD INFORMATION ORDER AND INTERNATIONAL ORDER
1981/82, FALL
J OF INTERNAT AFFAIRS
32, N2 179-188

INTERNAT INFO POLICY POLIT
ECON

36 ARMES, G.
THE KEY TO TECHNOLOGY (SATELLITE COMMUNICATION)
1984, JULY
SATELLITE COMMUNICATIONS
8 N7 P27-30

SAT TECH

37 ARND, GUY
TELECOMMUNICATION: IN THE VANGUARD OF PROGRESS(RECENT ADVANCE IN INFORMATION TECHNOLOGY AND SOME RECENT PROJECTS IN AFRICA
1982, MAY/JNE
MODERN AF
N6 P39

AF INFO TECH

38 ARONSON, S.H.
THE SOCIOLOGY OF THE TELEPHONE
1971
J OF COMPARATIVE SOCIOLOGY
12 P153-167

SOC TELEPHONE

39 ARTLE, R. & AVEROUS, C.
THE TELEPHONE SYSTEM AS A PUBLIC GOOD: STATIC & DYNAMIC ASPECTS
1973
BELL J OF ECON
4

TELEPHONE POLICY SOC EVAL
POLICY US

40 ASTRAIN, SANTIAGO
INTELSAT AND COMMUNICATIONS DEVELOPMENT: PAST, PRESENT AND FUTURE
1983 MAY 17
SPEECH/SOCIETY FOR INTERNATIONAL DEVELOPMENT/UNITED NATIONS/NEW YORK

INTELSAT SAT EVAL

41 ASTRAIN, SANTIAGO
INTELSAT AND TELECOMMUNICATION FOR THE DEVELOPING COUNTRIES EUROCOM
1977, MAY

INTELSAT TELECOMM SAT

42 ASTRAIN, SANTIAGO
SATELLITE COMMUNICATIONS: INTELSAT AND GLOBAL PATTERNS
1983, OCTOBER
SPACE COMMUNICATION AND BROADCASTING
1 N3 P253-260

WORLD TELECOMM SOC ECON
TRANSP INTELSAT COST RURAL
REMOTE

43 ASTRAIN, SANTIAGO
TELECOMMUNICATION AND THE ECONOMICS OF COMMUNICATION
SATELLITES
1981
ACTA ASTRONAUTICA
8 N11-1

ECON SAT INTELSAT TELECOMM

44 ASTRAIN, SANTIAGO
THE ROLE OF THE INTELSAT SYSTEM IN MEETING INTERNATIONAL &
NATIONAL TELECOMMUNICATION NEEDS
1978
PAPER AT INTERNAT SAT COMM SEMINAR, LIMA PERU

INTERNAT TELECOMM EVAL PLAN

45 AUSTINE, GORDON P.
ESTABLISHING A TELECOMMUNICATIONS TRAINING CENTER IN A DEVELOPING
COUNTRY
1981, JUNE
TELECOMMUNICATION JOURNAL
48

TELECOMM TRAIN

46 BACOLD, G.
THE WORKERS' RADIO: AN EXPERIMENT ON A SECTORAL MEDIUM
1980
MEDIA ASIA
7 N2 P91-94

PHILIPPINES RADIO

47 BADE, ANDRES, & BOWER, LINDA
TELECOMMUNICATION IN LATIN AMERICA: A CHANGING MARKETPLACE
1984, JULY 2
TELEPHONY
207 N1 P30-4

TELECOMM L AMER BUSI EVAL

48 BAGGALEY, J. AND SMITH, K.
FORMATIVE RESEARCH IN RURAL EDUCATION
MEDIA ASIA
9 N4 P196-202

RADIO RES EDUC RURAL
NEWFOUNDLAND TV

49 BAILEY, R.
THE MIDDLE EAST: A DECADE OF PROGRESS (TELECOMM DEVELOPMENT)
1984, AUG
COMM INT (GB)
11 N8 P33-42

M EAST TELECOMM HIST ECON

50 BAIR , I. & LEONHARD, J.
A DOMESTIC SATELLITE COMMUNICATION SYSTEM FOR ALGERIA
1975
PAPER AT IEEE CONF ON COMM

SAT ALGERIA AF PLAN

51 BALAKRISHNAN, R.
AIBD (ASIA PACIFIC INSTITUTE FOR BROADCASTING DEVELOPMENT)
1981, DECEMBER
MEDIA IN EDUCATION AND DEVELOPMENT
14 N4 P164-166

ASIA PAC BROADCAST TRAIN

52 BALDERSTON, MICHAEL
SATELLITE COMMUNICATIONS FOR EDUCATION IN THE SOUTH PACIFIC
1979, APRIL
EDUCATIONAL TECHNOLOGY
19 N4 P26-30

REMOTE SAT SOUTH PAC PAC
EDUC

53 BALL, D.W
TOWARD A SOCIOLOGY OF TELEPHONES AND TELEPHONERS
IN M.TRUZZI (ED) SOCIOLOGY IN EVERYDAY LIFE
1968

TELEPHONE SOC

54 BANDE, A., AND BOWER, L.
TELECOMMUNICATION IN LATIN AMERICA
1984, JULY
TELEPHONY
207 N1 30-4

TELECOMM L AMER TELEPHONE

55 BARBER, R.J.
TELECOMMUNICATION FACILITIES IN THE PACIFIC -- THE COMING
ABUNDANCE
1984, FEB
SIGNAL (USA)
38 N6

TELECOMM PAC DIGITAL FIBER
INTELSAT PLAN

56 BARK, ERIK
APPROPRIATE INFORMATION TECHNOLOGY:A CROSS-CULTURAL PERSPECTIVE
1982, OCT-DEC
UNESCO J OF INFO SCIENCE
4 N4
ERIC EJ276802

TECH EVAL COMPUTER INFO
TELECOMM

57 BARNHART, RAY & BARNHART, CAROL
CHIPPING AWAY AT RURAL SCHOOL PROBLEMS: THE ALASKAN EXPERIENCE WITH EDUCATIONAL TECHNOLOGY
1983, DEC
PHI DELTA KAPPAN
65 N4 P274-78

COMPUTER EDUC TECH ALASKA
RURAL TELECOMM COST

58 BASHSHUR, RASHID
TECHNOLOGY SERVES THE PEOPLE:THE STORY OF A COOPERATIVE TELEMEDICINE PROJECT BY NASA, THE INDIAN HEALTH SERVICE AND THE PAPAGO PEOPLE
1983
PROCEEDINGS OF TELECOMMUNICATION POLICY RESEARCH CONFERENCE
ABLEX: NORWOOD, N.J.

TELECOMM US RURAL HEALTH
NASA N AMER

59 BATES, T. AND ROBINSON, J.
EVALUATING EDUCATIONAL TV AND RADIO
1977
OPEN UNIV PRESS: ENGLAND

EUR EVAL EDUC TV
RADIO UK

60 BATES, TONY
APPLYING NEW TECHNOLOGY TO DISTANCE EDUCATION: A CASE STUDY FROM OPEN UNIVERSITY OF DIFFICULTIES IN INNOVATION
1980, SEPTEMBER
EDUCATIONAL BROADCASTING JOURNAL
P110-114

TECH EDUC REMOTE UK
RADIO

61 BATRA, V.
ZERO TO FIFTY IN FIVE YEARS (ALASKAN TELEPHONE NETWORK)
1984, MAY 28
TELEPHONY
206 N22

ALASKA TELECOMM TELEPHONE HIST

62 BATTU, DANIEL PIERRE AND ROSE, JOHN B.
TELECOMMUNICATION SERVICES FOR THE TRANSFER OF INFORMATION AND DATA: A CASE STUDY IN INDONESIA
ED246858
ERIC DOC REPRO SERV

INDONESIA TELECOMM COMPUTER INFO

63 BAUGHCUM, A., FAULHABER, G., VOIGHT, M.(ED)
TELECOMMUNICATION ACCESS & PUBLIC POLICY
1984
ABLEX PUB
TELECOMM

POLICY POLIT SOC ECON

64 BEAL, J.C.
A MULTIDISCIPLINARY INVESTIGATION OF THE APPLICATION OF DYNAMIC MODELING TO THE STUDY OF TELECOMM DEVELOPMENT IN CANADA
1976
IEEE TRANSACTIONS ON ENGIN. MGT
23 N2 P89-92

CANADA TELECOMM MODEL EVAL

65 BEAN, CLIFFORD
WORLDWIDE MOBILE TELECOMMUNICATIONS DEVELOPMENTS.
INTERNATIONAL TELECOMMUNICATION, LANDIS, K. (ED.)
1982
P87-101
D.C. HEATH: LEXINGTON, MA

WORLD TELECOMM INTERNAT

66 BEBEE, E.L. AND GILLING, E.W.J.
TELECOMMUNICATIONS AND ECONOMIC DEVELOPMENT: A MODEL FOR PLANNING AND POLICY MAKING
1976, AUGUST
TELECOMMUNICATION JOURNAL
43 N8 P537-542

TELECOMM ECON RES PLAN
POLICY

67 BECHER, E., (ED)
ON THE ECONOMICS AND FINANCE OF COMMUNICATION SERVICES: ECONOMIC EVALUATION FOR DECISION-MAKING
1974
MIMEO

ECON FIN EVAL

68 BECKER, ANN DE VANEY
A SURVEY AND EVALUATION: TELECONFERENCING
1978
JOURNAL OF COMMUNICATION
28 N3 P120-124

TELECONF EVAL RES

69 BECKER, JOERG
INFORMATION TECHNOLOGY AND A NEW INTERNATIONAL ORDER
1984
P24-40
LUND: STUDENTLITTERATUR

RURAL SAT TV SOC
CULT

70 BELKHAMZA, NOUR-EDDINE
DEVELOPMENT & IMPROVEMENT OF TELEPHONE NETWORK IN ALGERIA AS A COMMUNICATION & ENTERTAINMENT PROBLEM: ANALYSIS & MODELING
1982
DISS. ABSTRACT INTERNAT
43/11-A P3729

ALGERIA TELEPHONE EVAL HIST
MODEL

71 BELLA, I.& ROSO G.
A TECHNOECONOMIC ASSESSMENT FOR THE INTRODUCTION OF NEW SERVICES IN TELECOMMUNICATION NETWORKS
1982, AUG
CSELT RAPP TEC (ITALY)
10 N4 P235-42

TECH TELECOMM TECHNICAL EVALUATION
ECON

72 BELTRAN, LUIS RAMIRO
NATIONAL COMMUNICATION POLICIES IN LATIN AMERICA: A GLANCE AT THE FIRST STEPS
1976
INTERNAT CONFER ON COMM POLICY AND PLAN FOR DEV

POLICY TELECOMM MASS MEDIA L AMER
PERU VENEZUELA BRAZIL

73 BENDER, H.K.
TELECOMMUNICATION SYSTEMS PROJECTS IN DEVELOPING COUNTRIES
1980, MAY
NACHRICHTENTECH Z.(NTZ) (GERMANY)
33 N5 294-9

TELECOMM PLAN

74 BENNETT, A.M. ET AL.
TELEHEALTH HANDBOOK. A GUIDE TO TELECOMM TECHNOLOGY FOR RURAL HEALTH CARE
1978, MAY
REPT DHEW-PHS-79-3210; NCHSR-79-1

N AMER U.S. TELECOMM HEALTH

75 BENNETT, J.M. AND KALMAN, R.E.
COMPUTERS IN DEVELOPING NATIONS
1981
NORTH HOLLAND: AMSTERDAM

INFO COMPUTER MALAYSIA SAT
EDUC TRAIN MGT PAPUA NEW GUINEA
PHILIPPINES

76 BENNETT, JACK, AND HARLEY, JOHN
DOMSAT: AUSTRALIAN SATELLITE SERVICE
1982
PACIFIC ISLANDS COMMUNICATIONS JOURNAL
2 N1 P29-33

AUSTRALIA SAT

77 BENSTEAD, G.
USP SATELLITE COMMUNICATION PROJECT: REP FOR THE EXPERIMENTAL YEARS
1975-76
1977

PAC SATELLITE

78 BERRY, J.F.
COMMENTS ON THE CONTRIBUTION FO TELECOMM TO DEVELOPMENT
WITH PARTICULAR EMPHASIS ON FRANCE AND SPAIN
1981
INTERNATIONAL TELECOMMUNICATION UNION: GENEVA

FRANCE SPAIN TELECOMM ECON

79 BERRY, J.F.
SOME PROBLEMS & REQUIREMENTS OF EUROPEAN USERS OF TELECOMM
IN POLISHUK & O'BRYANT, EDS, TELECOMM & ECON DEVELOPMENT
1977
HORIZON HOUSE INTL: DEDHAM, MA

EUR TELECOMM ECON

80 BETH, C. RICHARD & DILMUS, JAMES
TECHNOLOGICAL PROGRESS IN LATIM AMERICA. JAMES STREET,
DILMUS, JAMES (EDS)
1979

INTERNAT TECH L AMER THEORY
ECON

81 BIDDLECOMBE, PETER
INTELCOM 1: INTEGRATED TELECOMMUNICATION FOR WEST AFRICA
1984, AUG 27
TELEPHONY
207 N10 P46-7

AF TELECOMM PLAN SAT

82 BLACKMAN, R.C.
TELECOMM FOR ECON DEVELOPMENT IN THE LOWER INCOME COUNTRIES
IN POLISHUK & O'BRYANT, EDS, TELECOMM & ECON DEVELOPMENT
1977
HORIZON HOUSE INTL: DEDHAM, MA

TELECOMM ECON SOC

83 BLACKMAN, R.C.
A ROLE FOR TELECOMMUNICATION IN THE ECONOMY OF AFRICA
& THE MIDDLE EAST
1977
UNPUBLISHED THESIS

AF M EAST TELECOMM ECON

84 BLANC, G.
THE IMPACTS OF TELECOMMUNICATIONS ON EMPLOYMENT
PARIS DEVELOPMENT CENTER
1981
ORGANIZATION FOR ECONOMIC COOPERATION AND DEVELOPMENT
ITU: GENEVA

TELECOMM ECON BUSI SOC

85 BLEVIS, B.C.
HISTORY, DEVELOPMENT AND FUTURE OF SATELLITE COMMUNICATION
IN CANADA
1979 MAY
CANADIAN J OF INFO SCIENCE
4 P41-51

CANADA SAT HIST RURAL
TECHNICAL

86 BLISS, A.G. & PHILLIS, B.C.
PROVIDING TELECOMM SERVICES TO REMOTE ABORIGINAL &
ISLANDER COMMUNITIES IN AUSTRALIA
1985
IN TELECOMM FOR PAC DEVELOPMENT, WEDMEYER, D.J. (ED)

AUSTRALIA TELECOMM TELEPHONE REMOTE

87 BLOCK, C., CHRISTENSEN, P., IMHOOF, M., SMITH, W.
RADIO IN THE CLASSROOM: TEN YEARS OF PROVEN SUCCESS
1984, SEPT

RADIO EDUC RES WORLD

88 BLOCK, C., GOLSCHMIDT, D., HAFID, A., LALOR, G. AND
VELASQUEZ, A.
SATELLITE TELECOMMS IN DEVMT: THE AID RURAL SATELLITE
PROGRAM AND ITS PILOT PROJECTS IN INDONESIA, PERU & CARIBBEAN
1984
PAC TELECOMM COUNCIL: HONOLULU, HI

INDONESIA CARIB PERU SAT
TECH FIN RURAL AID

89 BLOCK, CLIFFORD AND CASEY-STAHMER, ANNA
SATELLITE PROGRAM TO AID RURAL DEVELOPMENT
1981, JUNE
UPLINK
P1-2

SAT RURAL

90 BLOCK, CLIFFORD H.
BILATERAL AID IN COMMUNICATIONS- THE U.S. AGENCY FOR INTERNATIONAL
DEVELOPMENT
1985, SPRING
JOURNAL OF COMMUNICATION
P27-38

US INTERNAT FIN

91 BLOCK, CLIFFORD H.
MATCHING EDUCATIONAL NEEDS WITH AVAILABLE TECHNOLOGY: WHAT IS HAPPENING IN THE REST OF THE WORLD

TV SAT INTERNAT EDUAID
RURAL

92 BLOCK, CLIFFORD H.
PROMISING STEP AT ACAPULCO: A US VIEW
1982
JOURNAL OF COMMUNICATION
32 N3 P60-70

US POLICY N AMER INTERNAT

93 BLUMENTHAL, S., MCNEILL, D.
COMBINED QUARTERLY TECHNOLOGY REP # 30. PLURIBUS SATELLITE IMP DEVELOP MOBILE ACCESS TERMINAL NETWORK
1983, SEPT
AD-A132 690/9
BBN-5408
NTIS

TECH SAT REMOTE TECHNICAL
DATA

94 BLUTH, B.J.
SOCIAL BENEFITS FOR DEVELOPING COUNTRIES
1981
ADVANCES IN EARTH ORIENTED APPLICATIONS OF SPACE TECH
1, N4, 223-228

SAT SOC RURAL POLICY

95 BLYTH, MARY
THE CHANGING ROLE OF THE TELEPHONE: ITS IMPACT ON BUSINESS EDUCATION
1985, FEB
J OF BUSI EDUC
60 N5 P194-97

US TELEPHONE ECON EDUC
BUSI

96 BODECHTEL, J., BODECHTEL, E., ET AL.
ECON ASSESSMENT OF A EUROPEAN REMOTE SENSING SATELLITE SYSTEM FOR AGRICULTURAL APPLICATIONS V1 & 2
1981 JAN
N82-32832/9 & N82-32832/1

EUR SAT DATA REMOTE
ECON AGRIC

97 BOETTINGER, H.M.
DOCTRINE, DECISION & DEVELOPMENT IN INTERNATIONAL TELECOMMUNICATION
1984, JUNE
TELECOMM POLICY (GB)
8 N2 P89-92

POLICY TELECOMM INTERNAT

98 BONILLA, A.
RURAL TELECOMMUNICATIONS IN COSTA RICA. IN TELECOMMUNICATIONS AND ECONOMIC DEVELOPMENT, POLISHUK, P. AND O'BRYANT, M. (EDS.)
1977
HORIZON HOUSE INTERNATIONAL: DEDHAM, MA

COSTA RICA TELECOMM RURAL L AMER

99 BORDENAVE, JUAN E. DIAZ
COMMUNICATIONS AND RURAL DEVELOPMENT
1977
UNESCO PRESS: PARIS

RURAL

100 BORSOS, CHARLES
WORLD NETWORK DEVELOPMENT: THE CRITICAL POINTS
1984, OCT 22
TELEPHONY
P25-29

WORLD

101 BORTNICK, J.
INTERNATIONAL INFORMATION FLOW: THE DEVELOPING WORLD PERSPECTIVE
1981, SUMMER
CORNELL INTERNAT LAW J
14, N2, 333-356

INFO INTERNAT POLICY ECON
SOC POLIT BRAZIL

102 BORTNICK, J.
INTERNATIONAL TELECOMMUNICATION & INFORMATION POLICY
SELECTED ISSUES FOR 1980S
1983
GPO ITEM 1039-A 1039 (MICROFICHE)

INTERNAT TELECOMM POLICY INFO
POLICY

103 BOYD, PAUL D.
CAUSES AND CURES OF COMMUNICATION NEGLECT IN DEVELOPMENT PLANNING
1975 MAR
EDUCATIONAL BROADCASTING INTERNATIONAL
8 N1

EDUC EVAL

104 BRACE, JUDY
SELECTED COMMUNICATIONS TRAINING INSTITUTIONS FOR DEVELOPING COUNTRIES PERSONNEL
1983, MAY
CLEARINGHOUSE ON DEVELOPMENT COMMUNICATIONS/ WASHINGTON, D.C.

TRAIN

105 BRADFORD, D.E
JOINT PRODUCTS, COLLECTIVE GOODS AND EXTERNAL EFFECTS: COMMENT
1971
JOURNAL OF POLIT ECONOMY
79 P1119-1128

106 BRANCO, CASTELLO
BRAZIL AND THE INFORMATION SOCIETIES OF THE TWENTY-FIRST CENTURY
1982
COMM & THE FUTURE: PROSPECTS, PROMISES, AND PROBLEMS
P7-11
WORLD FUTURE SOCIETY: BETHESDA, MD

INFO POLICY TECH BRAZIL

107 BRANDT COMMISSION
NORTH-SOUTH: A PROGRAM FOR SURVIVAL: REPORT OF THE INDEPENDENT COMM ON INTERNATIONAL DEVELOPMENT ISSUES
1980

INTERNAT ECON

108 BREKKA, L.
EDUCATIONAL TV DEVELOPMENT IN IRAN
1976, JULY
IEEE TRANSACTIONS ON COMM
24 N7 P755-62

IRAN TV EDUC PLAN
M EAST

109 BRIMMER K.W.
UNITED STATES TELECOMMUNICATION COMMON CARRIER POLICY
1982
ANNUAL REVIEW OF INFO SCIENCE AND TECH
17 P33-82

TELECOMM DEREG US POLICY

110 BRIMMER, K. W.
TELECOMMUNICATION REGULATIONS AND POLICY
1981, DEC
ONLINE REV (GB)
5 N6 P481-6

TELECOMM POLICY LEGAL REG
US

111 BRISKMAN, R.D.
DOMESTIC COMMUNICATION SATELLITES: SYSTEM CONSIDERATIONS FOR RURAL AREAS
1981, NOV
IEEE COMUN MAG
19 N6 26-33

SAT RURAL TECHNICAL TELEPHONE
TV ECON

112 BRISKMAN, R.D.
DOMESTIC SATELLITE SERVICES FOR RURAL AREAS
1984, MARCH
IEEE COMMUN MAG
22, N3, 35-8

SAT RURAL TECHNICAL

113 BRUNEL, L.
TELECOMMUNICATION, DES MACHINES ET DES HOMMES
1978
LES DOSSIERS DE QUEBEC SCIENCE

TELECOMM

114 BURKE, W.R., LONGDON, N.
THE ECONOMIC IMPACT OF AN OPERATIONAL EUROPEAN METEOROLOGICAL SATELLITE PROGRAM IN THE 1980'S
1980, DEC
N82-20020/5

SAT DATA EUR ECON

115 BURROWES, R.
MULTIPLE TIME-SERIES ANALYSIS OF NATION-LEVEL DATA
1970
COMPARATIVE POLIT STUDIES
2 P465-480

STAT INTERNAT ECON RES
MODEL

116 BURTON, J.M.
TELECOMMUNICATION TECHNOLOGY IN THE OUTBACK
1984
TELECOMM J AUST (AUSTRALIA)
34, N 3, 213-19

AUSTRALIA RURAL TELECOMM

117 BUTLER, D.
TELECOMMUNICATION AND SOCIETY - KEY POLICY ISSUES FOR THE FUTURE
1980, FEB 4
TELEPHONY
198 N5

TELECOMM SOC POLICY

118 BUTLER, RICHARD E.
EVERYBODY ON LINE: THE TELEPHONE AS A KEY TO DEVELOPMENT
1983
MASS MEDIA
10 N3 P144-149

TELEPHONE RURAL COST ECON
SOC

119 BUTLER, RICHARD E.
GLODOM: CHEAP TELEPHONE CALLS FOR THE THIRD WORLD
1983, NOVEMBER
INTERMEDIA
11 N6 P45-50
INTERNATIONAL INSTITUTE OF COMMUNICATION

TELECOMM TELEPHONE RURAL COST
SATE ECON

120 BUTLER, RICHARD E.
INTERNATIONAL COOPERATION IN HUMAN RESOURCES MANAGEMENT
1986
TELECOMMUNICATION JOURNAL
53 N4 P198-202

INTERNAT TELECOMM TRAIN TECH

121 BUTLER, RICHARD E.
SATELLITE COMMUNICATIONS REGULATORY FRAMEWORK AND APPLICATIONS FOR DEVELOPMENT
1985
SPACE COMMUNICATION AND BROADCASTING
P103-108

SAT REG

122 BUTLER, RICHARD E.
THE BAREFOOT MICROCHIP ON THE LINE: WORLD COMMUNICATIONS YEAR-HIGHLIGHTS BEARING ON RURAL DEVELOPMENT
1983, JAN-FEB
DEVELOPMENT FORUM
P7

RURAL

123 BUTTEDAHL, PAZ G.
COMMUNICATIONS TECHNOLOGY AND ADULT EDUCATION: CAN PARTICIPATION BE ENCOURAGED
1983, JUNE
LIFELONG LEARNING: THE ADULT YEARS
6 N10 P4-6

TELECOMM SAT EDUC

124 BYRNES, T.J.
NATIONAL STRATEGIES FOR THE DEVELOPMENT OF A TELECOMM INFRASTRUCTURE IN IRELAND
1982, NOV
AD-P001 469/6

IRELAND POLIT POLICY TECH
TELECOMM

125 BYSTROM, JOHN
PEACESAT
1978, JANUARY
WORLD HEALTH ORGANIZATION
P10-13

SAT

126 CACCOAIAMI, E.R., GARNER, W.B., SALAMOFF,S.B.
THE EMERGENCE OF SATELLITE SYSTEMS FOR RURAL COMMUNICATION
1984, NOV
PROC. IEEE
72 N11 1445-56

SAT TELECOMM RURAL

127 CALVANO, MICHAEL
INDONESIAN NON-TELEPHONIC SATELLITE APPLICATIONS
1986, FEB 9
ACADEMY FOR EDUCATIONAL DEVELOPMENT: WASHINGTON, D.C.

INDONESIA SAT

128 CAMPBELL, N.A., ET AL.
EVALUATION & DEVELOPMENT OF TECHNIQUES FOR CROP INVENTORY
IN THE WHEATBELT OF WESTERN AUSTRALIA USING SATELLITE DATA
1982, JUNE
AD-P002 038/8

AGRIC EVAL TELECOMM SAT
RURAL AUSTRALIA

129 CANAS, M., AND ANTONIO, F.
SYSTEM DEVELOPMENT PATTERNS AND POLICIES: THE CASE OF COSTA RICA. IN
TELECOMMUNICATIONS AND ECONOMIC DEVELOPMENT, POLISHUK AND O'BRYANT
1977
HORIZON HOUSE INTERNATIONAL: DEDHAM, MA

L AMER COSTA RICA POLICY

130 CANTOR, A., MILLER, C., MURPHY, E. AND LARISEY, L.
COMPARISON OF MEDIA FOR DISSEMINATION OF INFORMATION ABOUT CANCER
IN RURAL COMMUNITIES
1979, OCTOBER
SOUTHERN MEDICAL JOURNAL
72 N10

SOUTH CAROLINA MEDIA EDUC HEALTH

131 CARIM, EMER
COMPUTER PROGRAM CUTS ENGINEERING COST
1985
THIRD WORLD DEVELOPMENT
2 P135

COMPUTER COST NIGERIA

132 CARREON, C.S.
THE REQUIREMENTS OF DEVELOPING COUNTRIES
1976
TELECOMM JOURNAL
43 P 124-129

TELECOMM

133 CARREON, C.S.
TELECOMMUNICATION, A RIGHT AND NOT A PRIVILEGE
1983
WORLD TELECOMMUNICATION FORUM: PART 1
INTERNATIONAL TELECOMMUNICATION UNION: GENEVA

TELECOMM POLICY SOC

134 CARREON, GEN. CAFERINO S.
THE ASIAN AND PACIFIC TELECOMMUNICATION NETWORK
1979, SEP 19-21
PAPER/ THIRD WORLD TELECOMMUNICATIONS FORUM
INTERNATIONAL TELECOMMUNICATION UNION: SWITZERLAND

ASIA PAC TELECOMM SAT
CABLE ITU

135 CARRILLO GAMBOA, E.
MEXICO PLANS AND IMPLEMENTS POLICY
1982, FEB
TELEPH ENG. AND MGT
86 N4 P57-9

MEXICO L AMER TELECOMM POLICY
SAT

136 CARTER, LAWRENCE J.
ONE VILLAGE, ONE RADIO IN RURAL PAPUA NEW GUINEA
1981, DECEMBER
TELEPHONY
20 N26 P62-69

PAPUA NEW GUINEA RADIO RURAL REMOTE

137 CARUSO, A.
ECONOMIC VICTIMS IN SATELLITE TELECOMMUNICATION
1984, AUG
TELECOMM
18 N8

SAT TELECOMM ECON EVAL

138 CASEY-STAHMER, ANNA
APPLICATIONS PLANNING, A DEVELOPMENT PROCESS
1981, JUNE
UPLINK
P4,6

PLAN

139 CASEY-STAHMER, ANNA
FINAL REPORT ON STUDY OF TWO-WAY RADIO IN MEDEX RURAL HEALTH PROJECT IN GUYANA
1979, MAY 15
ACADEMY FOR EDUCATIONAL DEVELOPMENT: WASHINGTON, D.C.

GUYANA 2-WAY RADIO HEALTH

140 CASEY-STAHMER, ANNA
SATELLITE PROJECT MANAGEMENT PAPER
1979, DEC 19
REPORT/ ACADEMY FOR EDUCATIONAL DEVELOPMENT/ WASHINGTON, D.C.
ACADEMY FOR EDUCATIONAL DEVELOPMENT: WASHINGTON, D.C.

SAT

141 CASEY-STAHMER, ANNA
THE ERA OF EXPERIMENTAL SATELLITES: WHERE TO GO FROM HERE?
1979, AUTUMN
JOURNAL OF COMMUNICATION
29, N4, 137-144

SAT RURAL PLAN EVAL
SOC ECON POLICY

142 CASEY-STAHMER, ANNA
THE RURAL COMMUNICATIONS SERVICES PILOT PROJECT IN PERU
1979, DECEMBER
ACADEMY FOR EDUCATIONAL DEVELOPMENT: WASHINGTON, D.C.

PERU L AMER RURAL

143 CASEY-STAHMER, ANNA AND LAUFFER, SANDRA
A NEW ERA OF SATELLITE COMMUNICATIONS: CAN EDUCATION
BENEFIT FROM IT?
1982, OCT
CONF PAPER/ INTL. CONF OF INTL. COUNC. FOR EDUC MEDIA

SAT WORLD EDUC

144 CASEY-STAHMER, ANNA, AND GOLDSCHMIDT, DOUGLAS
WAVES OF INFORMATION: USING RADIO AND COMMUNICATION SATELLITE
TECHNOLOGIES FOR RURAL DEVELOPMENT
1983, FEB 23-24
REPORT/ CENTER OF INTERNATIONAL CONFERENCE/ PARIS, FRANCE
DEVELOPMENT FORUM

SAT RADIO INFO RURAL

145 CASEY-STAHMER, ANNA, ET AL.
CRITICAL ASPECTS OF THE ESTABLISHMENT AND UTILIZATION OF SATELLITE
SERVICES FOR AFRICA
1984, APRIL
REPORT/ ACADEMY FOR EDUCATIONAL DEVELOPMENT
EUROPEAN SPACE AGENCY CONTRACT #5536/83/F/RD

AF SAT TELEPHONE TV
ECON RURAL EDUC HEALTH
DATA TRAIN

146 CHAMOUX, J.P.
INTERNATIONAL TELECOMMUNICATION POLICIES: A EUROPEAN PERSPECTIVE
1981, JAN
TRANSNATL DATA REP (NETHERLANDS)
4 N1

EUR INTERNAT POLICY FRANCE
TELECOMM

147 CHANDER, RAMESH, AND KARNIK, KIRAN
PLANNING FOR SATELLITE BROADCASTING: THE INDIAN INSTRUCTIONAL
TELEVISION EXPERIMENT
1976
UNESCO: PARIS

INDIA SAT TV EDUC
PLAN

148 CHAPLIN,J., FROMM, H.H., ROSETTI, C.
BROADCAST OF SOUND RADIO PROGRAMMES BY SATELLITE
DIRECT TO PORTABLE & VEHICLE RECEIVERS: A PROPOSAL FOR AFRICA
1984, MARCH
SPACE COMM & BROADCAST (NETHERLANDS)
2 N1 P3-11

RADIO BROADCAST

149 CHAPPEL, G.A.
TELECOMMUNICATION: A CHALLENGE AND AN OPPORTUNITY FOR THE
DEVELOPING WORLD
1984, FEB
RADIO AND ELECTRON. ENG (GB)
54 N2 55-7

WORLD TELECOMM ITU FIN

150 CHARLES, JEFFERSON O.
SATELLITE USE IN THE ENGLISH-SPEAKING CARIBBEAN: TOWARDS A MORE
INTEGRATED MODEL
1983
DISSERTATION
UNIVERSITY MICROFILMS INTERNATIONAL: ANN ARBOR, MI

CARIB SAT EVAL

151 CHARYK, J.V.
DEVELOPMENT OF INFORMATION AND TELECOMMUNICATION SYSTEMS
1984
TELEMATICS AND INFORMATICS
1 N1 P3-12

TELECOMM INFO ECON TECH

152 CHASIA, H.
CHOICE OF TECHNOLOGY FOR RURAL TELECOMMUNICATION IN DEVELOPING COUNTRIES
1976
IEEE TRANSACTIONS ON COMMUNICATIONS
24 N7 732-6

RURAL TELECOMM ITU TECH

153 CHAVES, N.
BRAZIL-MOVING INTO ITS OWN (SATELLITE COMMUNICATION DEVELOPMENT)
1984, JULY
SATELLITE COMMUNICATION
8 N7 P24-5

BRAZIL SAT ECON

154 CHECCHI AND CO.
SATELLITE AND TERRESTRIAL TELECOMMUNICATION DEVELOPMENT
1979
NASN-CR-107151; CONTRACT NASA 1357

AF SAT TELECOMM ECON

155 CHEN, P.C.
THE GROWTH AND DEVELOPMENT PLANS FOR TELECOMM IN TAIWAN
MAY, 1981
TELECOMM
15 N6

TAIWAN TELECOMM ASIA PLAN

156 CHENG, GEORGE CHIWO
AN EXPERIMENT OF APPROACHING THE CHINA MARKET
(TELECOMM, INFORMATION TECHNOLOGY, ECONOMICS)
1984
DISSERTATION ABSTRACTS INTERNAT
45/12-A P3682

TELECOMM INFO TECH ECON
CHINA ASIA

157 CHERRY, C.
THE TELEPHONE SYSTEM: CREATOR OF MOBILITY & SOCIETY
IN THE SOCIAL IMPACT OF THE TELEPHONE, POOL, I.(ED)
1977
MIT PRESS: CAMBRIDGE, MA.

TELEPHONE SOC

158 CHERRY, C.
WORLD COMMUNICATION: THREAT OR PROMISE?
1971
WILEY-INTERSCIENCE: NEW YORK

WORLD EVAL

159 CHIPMAN, RALPH
THE NEW COMPETITOR IN INTERNATIONAL TELECOMMUNICATIONS
1985, APRIL
NATIONAL DEVELOPMENT ASIA
P50-56

INTERNAT TELECOMM

160 CHIRKOV, B.
BEAMING TO ELEVEN TIME ZONES (US SATELLITE TV DEVELOPMENT)
1984 DEC
CABLE & SAT EUR (GB)
N12, 13-15

SAT US HIST

161 CHOI, SOON DAL
STRATEGIES ON DEVELOPMENT IN TELECOMMUNICATIONS FOR THIRD WORLD COUNTRIES -- DEVELOPMENT PROCESS & STRATEGIES IN KOREA
1983
TELECOMMUNICATIONS FORUM
INTERNATIONAL TELECOMMUNICATION UNION: GENEVA

KOREA TELECOMM

162 CHOWDARY, T.H
TELEPHONE GROWTH & NATIONAL DEVELOPMENT (IN INDIA)
1983, DEC
TELECOMM (INDIA)
33 N2 P25-35

TELEPHONE ECON INDIA

163 CHOWDARY, T.H.
TELECOMMUNICATION IN THE RURAL AREAS OF ANDHRA PRADESH
1984, JUNE
IETE TECH REV (INDIA)
1 N1 P7-12

TELECOMM RURAL INDIA

164 CHU, G.C., RAHIM, S. A. AND KINCAID, D.L. (EDS.)
COMMUNICATION FOR GROUP TRANSFORMATION IN DEVELOPMENT
1976
COMMUNICATION MONOGRAPH NO. 2. EAST-WEST CENTER
EAST-WEST CENTER: HONOLULU

CHINA SOC THEORY RES

165 CHU, GODWIN C., AND ALFIAN
PROGRAMMING FOR DEVELOPMENT IN INDONESIA
1980, AUTUMN
JOURNAL OF COMMUNICATION
30 N4 P50-65

INDONESIA

166 CHU, GODWIN, RAHIM, SYED & KINCAID, LAWRENCE (EDS)
INSTITUTIONAL EXPLORATION IN COMMUNICATION TECHNOLOGY
1978
EAST-WEST CENTER: HONOLULU, HAWAII

TECH POLICY POLIT CULT

167 CICONE, R., CIRST, E., METZLER, M., PARRIS, T.
DEVELOPMENT, IMPLEMENTATION & EVALUATION OF SATELLITE-AIDED
AGRICULTURAL MONITORING SYSTEMS-FINAL REPORT
1982, NOV
N83-34398/8

AGRIC SAT TELECOMM EVAL

168 CIFFI-REVILLA, CLAUDIO & MERRIT, RICHARD L.
COMMUNICATIONS RESEARCH AND THE NEW WORLD INFORMATION ORDER
1981/82, FALL/W
J OF INTERNAT AFFAIRS
35, N2, P225-24

RES INTERNAT THEORY SOC
POLIT

169 CLARK, DAVID & UNWIN, KATHRYN I.
INFO SERVICES IN RURAL AREAS: PROSPECTS
1980
GEO ABSTRACTS

ENGLAND INFO RURAL EVAL

170 CLARK, DAVID & UNWIN, KATHRYN I.
TELECOMM & TRAVEL: A REJOINER TO C.E. MILLER
1982, APRIL
REGIONAL STUDIES
16 N2 139-140

TRANS TELECOMM

171 CLARK, DAVID & UNWIN, KATHRYN I.
TELECOMMUNICATION & TRAVEL: POTENTIAL IMPACT IN RURAL AREAS
1981
REGIONAL STUDIES
15 N1 P47-56

TELECOMM RURAL EVAL TRANSP

172 CLARKE, ARTHUR C.
NEW COMMUNICATIONS TECHNOLOGIES AND THE DEVELOPING WORLD
1983, FEB 23-24
DEVELOPMENT FORUM
UNITED NATIONS: PARIS FRANCE

TELECOMM TECH SAT BROADCAST

173 CLARKE, ARTHUR C.
THIRD WORLD 2001
1981, NOVEMBER
SOUTH THE THIRD WORLD MAGAZINE
23 P13-15

TECH

174 CLARKE, D., SPOONER, H.J., LAXDAL, O.E., JOHNSON, D.I.
TELEPHONE IN CONTINUING MEDICAL EDUCATION -- SASKATCHEWAN EXPERIENCE
1979
MEDICAL TEACHER
N6 P298

EDUC TELEPHONE TELECOMM SASKATCHEWAN
CANADA RURAL

175 CLARKE, D.G. & LAUFENBERG, W.
THE ROLE OF TELECOMMUNICATION IN ECONOMIC DEVELOPMENT WITH SPECIAL REFERENCE TO RURAL SUB-SAHARA AFRICA
1981
INTERNATIONAL TELECOMMUNICATION UNION: GENEVA

RURAL AF TELECOMM ECON

176 CLEEVELY, D.
REGIONAL STRUCTURE & TELECOMMUNICATION PLANNNING
1984, JUNE
TELECOMM POLICY (GB)
8 N2 P149-61

TELECOMM PLAN

177 CLEEVELY, D., WALSHAM, G
TELECOMM MODELS: PLANNING FOR REGIONAL DEVELOPMENT IN LDC'S
1980, JUNE
TELECOMM POLICY (GB)
4 N2 P108-18

TELECOMM MODEL PLAN RURAL

178 CLEVELAND, HARLAN
THE FUTURE OF THE PACIFIC BASIN: A KEYNOTE ADDRESS
ED235091
ERIC DOC REPRO SERV

PAC TELECOMM SAT

179 CLIPPINGER, J.H. & FAIN, S.B.
PLANNING AND EVALUATING TELECOMMUNICATION DEMONSTRATIONS PROJECTS
AND ASSESSING THE COSTS OF TELECOMMUNICATION DEMONSTRATION PROJECTS
FINAL REP 146-03
ED237057
ERIC DOC REPRO SERV

TELECOMM PLAN ECON EVAL

180 CLIPPINGER, JOHN H.
A FRAMEWORK FOR NEEDS ASSESSMENT IN COMMUNICATIONS
DEVELOPMENT
1980, SEPT
TELECOMM POLICY
4, N3, P208-214

RES PLAN INFO POLICY
INTERNAT

181 CLIPPINGER, JOHN H.
CAN COMMUNICATION DEVELOPMENT BENEFIT THE THIRD WORLD?
1976, SEPTEMBER
TELECOMMUNICATION POLICY
1 P298-304

THEORY EVAL TECH

182 CLIPPINGER, JOHN H.
THE HIDDEN AGENDA
1979, WINTER
JOURNAL OF COMMUNICATION
29, N1 197-203

INTERNAT SAT POLICY INFO
SPECT ITU

183 CLIPPINGER, JOHN H.
WHO GAINS BY COMMUNICATION DEVELOPMENT? STUDIES IN
INFORMATION TECHNOLOGIES IN DEVELOPING COUNTRIES
1976, JAN
CENTER FOR INFORMATION POLICY RESEARCH
HARVARD UNIVERSITY: CAMBRIDGE, MA

ALGERIA EL SALVADOR INFO TECH
POLIT ECON SOC CULT
TV EDUC

184 CLIPPINGER, JOHN H., AND FAIN, SANFORD B.
EVALUATION OF TELECOMMUNICATIONS DEMONSTRATIONS PROJECT AND
RECOMMENDATIONS TO THE NEW TELECOMMUNICATIONS DEMONSTRATIONS PROGRAM
1980
REPORT/DEPARTMENT OF HEALTH

TELECOMM EVAL MODEL

185 CODDING, GEORGE A. JR.
DECISION MAKING IN THE ITU: THE NATURE AND ROLE OF
THE NEW THIRD WORLD MAJORITY
1980
PB82-104977

NTIS & NSF

ITU

186 CODDING, GEORGE A. JR. & RUTKOWSKI, ANTHONY M.
THE INTERNATIONAL TELECOMMUNICATION UNION IN A CHANGING WORLD
1982
ARTECH: DEDHAM, MA

ITU INTERNAT POLICY

187 COLINO, RICHARD R.
BRIDGING THE GAP IN CARIBBEAN COMMUNICATIONS: THE ROLE OF INTELSAT
1986, MAR 20
PAPER/ CONF OF THE CARIBBEAN ASSOC OF NATIONAL TELECOMM OPERATORS
NASSAU, BAHAMAS

CARIB INTELSAT SAT

188 COLINO, RICHARD R.
CLOSING THE GAPS IN A SHRINKING WORLD: INTELSAT AND SATELLITE COMMUNICATIONS IN THE DEVELOPING COUNTRIES
1984, OCTOBER
NEW DIRECTIONS IN SAT COMMUNICATIONS FOR NORTH AND SOUTH
P1-10
UNIVERSITY OF TEXAS, AUSTIN

INTELSAT SAT RURAL

189 COLLE, ROYAL D.
COMMUNICATION SYSTEMS AND THE NEW RURAL DEVELOPMENT STRATEGIES
1976, JULY
PAPER PRESENTED AT ADVANCED SUMMER SEMINAR ON COMM TECH

RURAL

200 CRUIEN, N. PAUTRAT,C.
AN ECONOMIC APPROACH TO TELECOMMUNICATION TARIFFS
1983, APRIL
TELECOMM JOURNAL (SWITZERLAND)
50 N4 P187-94

ECON TELEPHONE TELECOMM REG

201 CRUISE O'BRIEN, R. AND HELLEINER, G.K.
THE POLITICAL ECONOMY OF INFO IN A CHANGING INTERNAT ECON ORDER
IN COMM ECON AND DEVELOPMENT M. JUSWALLA & D.M. LAMBERTON (EDS)
1982

POLIT ECON INFO SOC
INTERNAT

202 CRUISE O'BRIEN, R., COOPER, R.E., PERKES, B. AND LUCAS, H.
COMMUNICATION INDICATORS AND INDICATORS OF SOCIOECONOMIC
DEVELOPMENT 1960-1970
1977
UNPUBLISHED STUDY/ UNIVERSITY OF SUSSEX: BRIGHTON, U.K.

SOC ECON STAT RES

203 CRUISE O'BRIEN, RITA
BALANCING OUT THE FLOW OF INFORMATION
1980, SUMMER
IRISH BROADCASTING REVIEW
N8, 23-26

INFO INTERNAT ECON POLIT
TELECOMM COMPUTER

204 CRUISE O'BRIEN, RITA
IS THERE A LINK BETWEEN TELECOMMUNICATION AND
DEVELOPMENT?
1979, JULY
INTERMEDIA
7, N4, 26-29

TELECOMM ECON INTERNAT PLAN
MASS MEDIA

205 CRUISE O'BRIEN, RITA
TECHNOLOGICAL FACTORS IN INTERNATIONAL COMMUNICATION
1980 APR
COMM
5:1

BROADCAST AF TECH WORLD

206 CUTLER, M.
USING RADIO TO TEACH LANGUAGE ARTS: PILOT PROGRAM IN KENYA
ADDRESSING PRIMARY NEEDS
1980
DEV COMM REPORT

RADIO KENYA EDUC AF

207 DANIEL, JOHN S., STRAND, M.A., THOMPSON, J.R. (EDS)
LEARNING AT A DISTANCE: A WORLD PERSPECTIVE
1982
ATHABASCA UNIVERSITY, INTERNAT. COUNCIL FOR CORRESP. ED: EDMONTON

COMPUTER POLICY TELECOMM ECON
EDUC REMOTE

208 DARMON, JACQUES
SHAPING THE FUTURE STRUCTURE OF GLOBAL COMMUNICATION
WORLD TELECOMMUNICATION FORUM. PART 1
1983
SYMPOSIUM ON POLICY, FINANCIAL AND ECONOMIC ASPECTS OF TELECOMM
INTERNATIONAL TELECOMMUNICATION UNION: SWITZERLAND

WORLD HIST TECH POLICY
SOC POLIT

209 DAS, J.
TELECOMMUNICATION AND SOCIAL OBJECTIVES
1980, NOV
JOURNAL OF INST. ELECTRON. AND TELECOMM ENG. (INDIA)
26 N11

INDIA TELECOMM SOC

210 DAUCHY, ERIC
INDONESIAN TRADITIONAL COMMUNICATIONS PATTERNS MASS MEDIA AND TELECOMMUNICATIONS: EMPHASIS ON PALAPA SATELLITE COMMUNICATIONS
1984
KATHOLEIKE UNIVESITEIT, LEUVEN

INDONESIA HIST EDUC TRANSP
POLIT MEDIA RADIO TV
TECH SAT

211 DAVID, E.C.
DZUP: A CASE STUDY
1982
MEDIA ASIA
9 N4 P206-212

PHILIPPINES RADIO EDUC

212 DAVIS, WARREN E.
HIGH TECHNOLOGY--THE DEVELOPING WORLD'S KEY TO PROSPERITY
1985
THIRD WORLD DEVELOPMENT
2 P179-180

TECH ECON

213 DAWIDZINK, M.B. AND PRESTON, H.F.
INTERNATIONAL COMMUNICATION NETWORK DEVELOPMENTS AND ECONOMIC
1981, JANUARY
TELECOMMUNICATION JOURNAL
8 N1 P19-31

WORLD CABLE SAT ECON

214 DEGNAN, K.E., DUNMORE, K.R., SARDELLA, V.
DIRECT BROADCAST SATELLITE: POLICIES, PROSPECTS & POTENTIAL COMPETITION
1981
GPO IIEM 126-D-3

SAT BROADCAST POLICY N AMER
US TV ECON

215 DELORME, G.A. & RAINER, J.M.
STUDY OF THE SITUATION AND THE NEEDS IN AFRICA RELATING TO METEROLOGICAL TELECOMMUNICATION
1978 MARCH
N79-11627/3 & REPT ESA-CR(P)-1091

SAT DATA AF RES

216 DEMMERT, J.P. AND WILKE, J.L.
THE LEARN/ALASKA NETWORKS: INSTRUCTIONAL TELECOMMUNICATIONS IN ALASKA
TELECOMMUNICATIONS IN ALASKA, ED. ROBERT WALP
1982
PACIFIC TELECOMMUNICATIONS COUNCIL: HONOLULU, HI

ALASKA EDUC TELECOMM SAT
RURAL N AMER

217 DHILLON, H.S., DOERMANN, A.C.
TELEMEDICINE AND RURAL PRIMARY HEALTH CARE - ANALYSIS OF THE IMPACT OF TELECOMMUNICATION TECHNOLOGY
1978
SOCIO-ECON PLANNING SCIENCES
12 N1 37-48

N AMER US HEALTH TELECOMM

218 DIAZ BORDENAVE, J.E.
LATIN AMERICA INITIATES A NEW APPROACH TO RURAL COMMUNICATION
1980, DECEMBER
EDUCATIONAL BROADCASTING INTERNATIONAL
13 N4 P163-167

L AMER RURAL AGRIC POLICY
INFO EDUC

219 DICKENSON, C.R.
TELECOMM IN THE DEVELOPING COUNTRIES: THE RELATION TO THE ECONOMY AND THE SOCIETY
IN POLISHUK & O'BRYANT, EDS, TELECOMM & ECON DEVELOPMENT
1977
HORIZON HOUSE INTL: DEDHAM, MA

TELECOMM ECON SOC

220 DICKENSON, C.R. AND SAUNDERS, R.J.
TELECOMMUNICATIONS: PRIORITY NEEDS FOR ECONOMIC DEVELOPMENT
1979
TELECOMMUNICATION JOURNAL
46 N9 P566-570

TELECOMM ECON POLICY

221 DICKS, D.
FROM DOG SLED TO DIAL TELEPHONE: A CULTURAL GAP?
JOURNAL OF COMMUNICATION
27 N4 P120-129

TELEPHONE CULT

222 DIEHL, JACKSON
INDEBTED, TANGLED PHONE SYSTEM BURDENS ARGENTINA
1984, JUNE 21
WASHINGTON POST
P31
WASHINGTON, D.C.

ARGENTINA TELEPHONE

223 DIJRE, DEMBELE SATA
RESEAU VILLAGEOIS ET UTILISATION DE LA VIDEO
1983, FEB 23-24
DEVELOPMENT FORUM
UNITED NATIONS: PARIS, FRANCE

VIDEO RURAL MALI

224 DINGELL, JOHN D.
TELECOMMUNICATION POLICY SHOULD BE PRAGMATIC
1985, APRIL 4
PUBLIC UTILITIES FORTNIGHTLY

TELECOMM POLICY POLIT EVAL

225 DIZARD, JOHN
THE REVOLUTION IN TELECOMMUNICATION FINANCE
1979
INSTITUTIONAL INVESTOR
13 P143-4

TELECOMM FIN US INVEST
TECH

226 DIZARD, WILSON P.
THE COMING INFO AGE: AN OVERVIEW OF TECHNOLOGY, ECONOMICS AND POLITICS
1982

INTERNAT SAT TELECOMM COMPUTER
POLICY US

227 DIZARD, WILSON P.
THE U.S. POSITION: DBS AND FREE FLOW
1980, SPRING
JOURNAL OF COMMUNICATION
30 N2 P157-167

US BROADCAST SAT CULT
SOC

228 DOBRISKI, A.S.
PLANNING CONSIDERATIONS FOR RURAL NETWORKS
1981, APRIL
TELECOMM J
48 N4 P189-91

PLAN RURAL

229 DORDICK, H.S.
TELECOMMUNICATIONS FOR TELECOMMUTING: PROSPECTS FOR ENERGY SAVINGS SUBSTITUTION
1980, MAY
PROCEEDINGS OF THE UNIV OF CA TRANSP PROGRAM
ANNUAL CONFER

TRANS TELECOMM COST

230 DORDICK, H.S.
TRANSBORDER DATA FLOWS: COMPETITION AND REGULATION
IN WORLD COMMUNICATIONS GERBNER, G. & SIEFER, M. (EDS)
1984
P 210-216
LONGMAN: NY

INFO DATA INTERNAT REG

231 DORDICK, H.S., BRADLEY, H.G., AND NANUS, B.
THE EMERGING NETWORK MARKETPLACE
1981
ABLEX: NORWOOD, NEW JERSEY

TELECOMM COMPUTER INFO NETWORK

232 DORDICK, HERBERT J.
HOW TO COPE WITH THE NEW INFORMATION AGE
1985
THIRD WORLD DEVELOPMENT
2 P190-192

INFO ECON

233 DORDICK, HERBERT S.
THE COMMUNICATIONS INDUSTRY AND ITS ROLE IN TODAY'S WORLD
1985
THIRD WORLD DEVELOPMENT
2 P34-38

WORLD TELECOMM

234 DORMOIS, M. AND GENSOLLEN. M.
LE MARCHE DU TELEPHONE
1976
ECONOMIE ET STATISTIQUE
78 P3-12

TELEPHONE STAT ECON EUR
FRANCE

235 DOUDS, M. JAY
THE RESHAPING OF AN INNOVATION. ASCN-THE LEARNING CHANNEL, 1974-1982
AN EXECUTIVE SUMMARY PREPARED FOR NATIONAL INSTITUTE OF EDUCATION
1982

EDUC SAT CABLE TV
REMOTE RURAL

236 DUBRET, F.
TELECOMMUNICATIONS AND THEIR IMPACT ON THE FISHING INDUSTRY
1981
INTERNATIONAL TELECOMMUNICATION UNION: GENEVA

TELECOMM BUSI ECON

237 DUNTT, COLIN
COMMUNICATING EDUCATION IN SOUTH AUSTRALIA
1983 DEC
MEDIA IN EDUC AND DEVELOPMENT
16 N4 PA62-4

EDUC TECH AUSTRALIA REMOTE
SAT INTERNAT

238 DURLING, G. W. & PFUND, E.T.
LOW COST COMMUNICATION SATELLITES FOR DEVELOPING COUNTRIES
1984
COMM SAT SYSTEMS CONF 10TH ORLANDO FL TECH PAPERS
A84-2525110-32
AMER INST OF AERONAUTICS & ASTRONAUTICS

SAT COST FIN RURAL
TECHNICAL

239 DUSEATH, D.H.J
THE DEVELOPMENT OF NEW ZEALAND RURAL TELECOMMUNICATION
1984 NOV
TRANS INST PROF ENG N.Z.
11, N3

TELECOMM NEW ZEALAND RURAL HIST

240 EAST WEST CENTER
COMMUNICATION POLICY & PLANNING FOR DEVELOPMENT: AN ANNOTATED BIBLIOGRAPHY
1976
EAST-WEST CENTER: HONOLULU, HAWAII

POLICY PLAN INTERNAT

241 EDGERTON, JERRY
TELECOMMUNICATION INVESTMENT
1983, JULY
MONEY
P62-64

TELECOMM FIN

242 EDUTEL COMM & DEVELOPMENT INC
SYSTEM STUDY OF TECHNOLOGY USEFUL FOR PROVIDING ECONOMICAL TELEPHONE SERVICE IN RURAL ALASKA
1978, MAY 31
PB-297 211/5, PB-297 210/7 OR REA/CR-79/01 & 02
I & II
REA & NTIS

TECH TELEPHONE RURAL ALASKA

243 EDWARD, RONALD S.
TELECOMMUNICATION POLICIES IN SEVENTEEN COUNTRIES: PROSPECTS FOR FUTURE COMPETITIVE ACCESS
ED258558
ERIC DOC REPRO SERV

WORLD TELECOMM POLICY ECON

244 EFTEKHARI, MOHAMMAD-REZA
FACSIMILE COMMUNICATION & ELECTRONIC MAIL SYSTEM BY SATELLITE IN IRAN & OTHER DEVELOPING NATIONS
1978
DISSERTATION ABSTRACTS INTERNAT
38/12-B P6072

IRAN SAT COMPUTER

245 EGBON, MIKE
WESTERN NIGERIA TV SERVICE -- OLDEST IN TROPICAL AFRICA
SUMMER 1983
JOURNALISM QUARTERLY
60 N2
ERIC # EJ28511

NIGERIA TV RES HIST
TELECOMM

246 EGER, JOHN M.
THE INTERNATIONAL INFORMATION WAR
1981, MARCH 18
COMPUTERWORLD EXTRA
5, N11A 103-119

INFO INTERNAT POLICY

247 ELLINGHAUS, W. AND FORRESTER, L.G.
A U.S. EFFORT TO PROVIDE A GLOBAL BALANCE- THE MAITLAND COMMISSION REPORT
1985, SPRING
JOURNAL OF COMMUNICATION
35 N2 P14-20

WORLD TELECOMM INFO TECH
SOC CULT ECON

248 ELLIS, L.W.
SOME SOCIO-ECONOMIC CONSEQUENCES OF THE SPREAD OF ELECTRONICS THROUGHOUT TELECOMMUNICATION
1980, SEPT
TELECOMM J (GB)
47 N9 572-7

TELECOMM SOC ECON

249 ELTON, M.C.J., LUCAS, W.A., AND CONRATH, D.W.
EVALUATING NEW TELECOMMUNICATIONS SERVICES
1978
NATO CONFERENCE SERIES. SERIES II: SYSTEMS SCIENCE
6
PLENUM PRESS: NEW YORK

TELECOMM EVAL RES TELECONF
V-TEX POLICY PLAN CANADA

250 ELY, DONALD P.
INFORMATION TECHNOLOGY IN EDUCATION: THE BEST OF ERIC
ED233706
ERIC DOC REPRO SERV

TELECONF COMPUTER TELECOMM SAT
EDUC

251 EMARD, JEAN-PAUL
INFORMATION & TELECOMMUNICATION-AN OVERVIEW OF ISSUES, TECHNOLOGY & APPLICATIONS
1981, JULY
SCI POLICY RES DIV CONGRESSIONAL RES SERV
97TH CONG. 1ST
U.S. HSE COMM ON SCI & TECH

N AMER US TELECOMM INFO
POLICY

252 ENCEL, S.
SOCIAL ASPECTS OF COMMUNICATION
1978
JOURNAL OF COMMUNICATIONS RESEARCH
4 N3 P271-287

SOC TELEPHONE

253 ENCEL, S.
SOCIAL ASPECTS OF COMMUNICATION
1975
IEEE TRANSACTIONS ON COMM
23 P1012-1018

SOC TELECOMM RES

254 ENGELBARDT, R.M.
ISLAND PARADISE GETS SYSTEM UPDATE (MICRONESIAN TEL'S TELEPHONE NETWORK DEVELOPMENT)
1984, SEPT 15
TEL ENG & MGT
88 N18 P104-5

MICRONESIA TELEPHONE HIST EVAL

255 ENGVALL, LARS
SOCIO-ECONOMIC STUDY OF THE USAGE OF THE RADIO, TELEVISION AND TELEPHONE SERVICES
1982
TELECOMMUNICATION JOURNAL
49 N11 P769-772

RADIO TV TELEPHONE SOC
ECON

256 EPSKAMP, C.P.
TRAINING AND RESEARCH WITH REGARD TO THE APPLICATION OF THE MASS MEDIA FOR EDUCATION AND INFORMATION IN LATIN AMERICA AND CARIBBEAN
1982
GAZETTE
29 N3 P180-196

L AMER CARIB MEDIA EDUC
INFO TRAIN RES RADIO
RURAL

257 EPSKAMP, C.P. AND BOEREN, A.J.J.M.
EDUCATION AND COMMUNICATION: A CONCEPTUAL FRAMEWORK
1979
CENTER FOR THE STUDY OF EDUC IN DEV COUNTRIES:THE HAGUE

EDUC EUR NETHERLANDS TELECOMM
INTERNAT TEHORY

258 ERNBERG, J.
MANUFACTURERS'/ SUPPLIERS' ROLE IN THE DEVELOPMENT OF TRAINING AND PERFORMANCE AIDS
1981
TELECOMMUNICATION JOURNAL
48 N8 P483-486

TELECOMM TRAIN

259 ERNES, BETH KREVITT
TRANSFER OF INFORMATION TECHNOLOGY TO LESS DEVELOPED COUNTRIES: A SYSTEMS APPROACH
1980, JANUARY
JOURNAL OF THE AMERICAN SOCIETY FOR INFO SCIENCE
P97-102

INFO TECH SOC ECON

260 ERNST, DIETER
TECHNOLOGY POLICY FOR SELF-RELIANCE: SOME MAJOR ISSUES
1981
INTERNAT SOC SCIENCE J
33, N3 466-480

TECH POLICY SOC

261 ESPINAL, C.
HONDUTEL: OUR CASE FOR NATIONAL TELECOMMUNICATIONS PLANNING. IN TELECOMMUNICATIONS AND ECONOMIC DEVELOPMENT, POLISHUK AND O'BRYANT
1977
PAPER/ INTELCOM 77/ ATLANTA, GEORGIA
HORIZON HOUSE INTERNATIONAL: DEDHAM, MA

TELECOMM PLAN

262 EUROPEAN SPACE AGENCY-PARIS
SATELLITE REMOTE SENSING. APPLICATIONS TO RURAL DISASTERS
1981
N81-27571/1
EUROPEAN SPACE AGENCY: PARIS

EUR SAT DATA REMOTE

263 FARR, J.P. & WALDRON, E.H.J.
ADMINISTRATION OF TRAFFIC FORECASTING FOR SPARSE RURAL TELEPHONE NETWORKS
1984
AUST TELECOMM RES (AUSTRALIA)
18 N1 P59-66

AUSTRALIA RURAL TELEPHONE TECH

264 FETT, J.
SITUATIONAL FACTORS AND PEASANTS' SEARCH FOR MARKET INFO
1975
JOURNALISM QUARTERLY
53 P429-435

INFO RURAL ECON

265 FILEP, R. T. AND PELTON, J. N.
EDUCATION AND COMMUNICATION SATELLITES: OPPORTUNITIES FOR OUTREACH
1983, OCT
34TH CONGRESS OF THE INTL AERONAUTICS FED. BUDAPEST, HUNGARY

EDUC SAT RURAL WORLD

266 FINCHMAN, A.G., DESAI, P., HALLIWELL, J., ET AL.
THE USE OF RADIO FOR THE IN-SERVICE CONTINUING EDUCATION OF RURAL PRIMARY HEALTH CARE PERSONNEL IN JAMAICA
1984
UNITED COOPERATIVE PRINTER LIMITED: KINGSTON, JAMAICA

JAMAICA RADIO EDUC HEALTH
RURAL

267 FLORES-HERNANDEZ, LORENZO
ECONOMIC CAPACITY EXPANSION PLAN: AN APPLICATION TO THE TELEPHONE INDUSTRY
1982
DISSERTATION ABSTRACTS INTERNAT
43/10-B P3324

ECON PLAN BUSI

268 FOISSEY, A.
BROADCASTING ASPECTS OF A PROJECTED AFRICAN REGIONAL SATELLITE TELECOMMUNICATIONS SYSTEM
1983, JULY
SPACE COMM AND BROADCAST (NETHERLANDS)
1 N2

AF SAT TELECOMM BROADCAST

269 FOOTE, D. R.
SATELLITE COMMUNICATION FOR RURAL HEALTH CARE IN ALASKA
1977
JOURNAL OF COMMUNICATION
27 N4 P173-182

ALASKA SAT HEALTH RURAL
COMPUTER EVAL

270 FOOTE, D.R.
THE UNIVERSITY OF THE WEST INDIES EXPLORES SATELLITE USE
1981, JUNE
UPLINK
P5

SAT EDUC CARIB

271 FOOTE, D.R., PARKER, E.B. & HUDSON, H.E.
HEALTH IN ALASKA: THE ATS-6 SAT BIOMEDICAL DEMO
1976
INSTITUTE FOR COMMUNICATION RESEARCH: STANFORD UNIVERSITY

HEALTH ALASKA SAT

272 FRASER, COLIN
SOME REALITIES IN THE APPLICATIONS OF COMMUNICATION TECHNOLOGY FOR RURAL DEVELOPMENT IN THE THIRD WORLD
1983, NOV 23-24
PAPER/ UNITED NATION DEVELOPMENT FORUM
CENTER FOR INTERNATIONAL CONFERENCE: PARIS

TECHNICAL RURAL

273 FRASER, J.M.
INTERFACES IN RURAL SATELLITE TELEPHONE SYSTEM
1982, NOVEMBER
PAPER/ GLOBECOM '82 (IEEE)/ MIAMI, FLORIDA
P606-610

SAT TELEPHONE RURAL

274 FREIRE, P.
EDUCATION FOR CRITICAL CONSCIOUSNESS
1973
SEABURY PRESS: NEW YORK

EDUC L AMER SOC POLICY

275 FREIRE, P.
PEDAGOGY OF THE OPPRESSED
(TRANSLATED BY MYRA BERGMAN RAMES)
1970
HERDER AND HERDER: NEW YORK

EDUC L AMER SOC POLICY

276 FREY, F.A.
COMMUNICATION & DEVELOPMENT, IN HANDBOOK OF COMMUNICATION
POOL,I., FRY, F., SCHRAMM, W.,MACCOBY, N. & PARKER, E.B.
1973

ECON SOC TELECOMM TV
SOC CULT

277 FRIEDEN, ROBERT
GETTING CLOSER TO THE SOURCE: NEW POLICIES FOR
INTERNATIONAL SATELLITE ACCESS
1985, APRIL
FEDERAL COMMUNICAITON LAW JOURNAL
37 N2 P293-323

POLICY SAT INTERNAT POLIT

278 FRIMAN, E.
THE NEW AGE OF MEDICAL TELEVISION NETWORKS: THE POTENTIAL
SATELLITE TRANSMISSION HOLDS FOR NETWORKING MEDICAL EDUCATION
1977, MAY
EDUCATIONAL AND INDUSTRIAL TELEVISION
19 N22

N AMER INDIANA SAT EDUC
HEALTH TRAIN

279 FRUITKIN, ARNOLD
COMMUNICATION TECHNOLOGY AND SOCIAL POLICY
IN GERBNER, G., GROSS, L., & MELODY, W.
1973

TECH SOC ECON POLICY
TELECOMM INTERNAT

280 GALL, ROBERT S.
KEYS TO THE WORLD: A BRIEF EXAMINATION OF MICROCOMPUTER TECHNOLOGY FOR TELECOMMUNICATION
ED258395
ERIC DOC REPRO SERV

TELECOMM COMPUTER RURAL EDUC

281 GANLEY, OSWALD & GANLEY, GLADYS
TO INFORM OR TO CONTROL? THE NEW COMMUNICATION NETWORKS.
1982
MC GRAW-HILL: NEW YORK

INFO TECH SAT INTERNAT
ECON POLIT POLICY US

282 GAOTENG, SHU
PRESENT STATUS AND FUTURE DEVELOPMENT OF CHINA'S TELECOMMUNICATIONS
1983
PAPER/ WORLD TELECOMMUNICATIONS FORUM
INTERNATIONAL TELECOMMUNICATIONS FORUM: SWITZERLAND

CHINA TELECOMM

283 GARCIA, R.B.
TELECOMMUNICATION AND INVESTMENT DECISIONS IN THE PHILIPPINES
MARCH 1984
TELECOMM POLICY (GB)
8 N1 P51-7

INVEST FIN PHILIPPINES TELEPHONE
POLICY PLAN

284 GARFINKLE, MAXIMO
ARGENTINA'S TELECOMMUNICATION MOVEMENT: PROJECT SOVEREIGNTY
1982, MARCH
TELECOMMUNICATIONS
16 N3 P81-86

ARGENTINA TELECOMM TELEPHONE TV
RADIO SAT

285 GARRIOTT, GARY
THE EVOLUTION OF A SATELLITE
1984, APRIL
VITA NEWS
P6-7

SAT

286 GEISER, U., SOMMER, M., HUMBEL, R., HAEFNER, H., AND ITTEN, K.I.
MULTISTAGE LAND USE MAPPING AND CHANGE MONITORING IN SRI-LANKA
1982, APR 20-21
PAPER/ PROCEEDINGS OF SATHE EARSEL-ESA SYMPOSIUM/ IGLS, AUSTRIA
P1-7

SRI-LANKA COMPUTER AGRIC PLAN

287 GELLERMAN, R.F.
APPROPRIATE TECHNOLOGY IN LATIN AMERICAN TELECOMM, IN TELECOMM AND ECONOMIC DEVELOPMENT, POLISHUK & O'BRYANT (EDS)
1977
PAPERS AT INTELCOM 77

L AMER TELECOMM ECON

288 GELLERMAN, ROBERT F.
TELECOMMUNICATIONS ACTIVITIES OF THE INTER-AMERICAN DEVELOPMENT BANK
1978
THE INTER-AMERICAN DEVELOPMENT BANK: WASHINGTON, D.C.

TELECOMM FIN

289 GELLERMAN,R.F. & LING, S.
LINKING ELECTRICITY WITH THE TELEPHONE DEMAND FORECAST:
A TECHNICAL NOTE
1976
IEEE TRANSACTIONS ON COMM
24 P322-325

TELEPHONE STAT PLAN TECH

290 GELLERMAN, ROBERT F.
MEASURING THE BENEFITS OF TELECOMMUNICATIONS
IN POLISHUK & O'BRYANT, EDS, TELECOMM & ECON DEVELOPMENT
1977
HORIZON HOUSE INTL: DEDHAM, MA

TELECOMM EVAL RES

291 GENERAL ACCOUNTING OFFICE
DEVELOPING A DOMESTIC COMMON CARRIER TELECOMMUNICATION POLICY:
WHAT ARE THE ISSUES
1979, JAN
PB-290 787/1 CED-79-18
GENERAL ACCOUNTING OFFICE, WASHINGTON, D.C.

US POLICY TELECOMM TELEPHONE
POLIT

292 GEORGE, V.V.
EDUCATIONAL MEDIA IN MALAYSIA
1982
MEDIA ASIA
9 N4 P213-217

MALAYSIA EDUC AV RADIO
TV RURAL FILM

293 GERBNER, GEORGE AND SIEFERT, MARSHA
WORLD COMMUNICATIONS - A HANDBOOK
1984
LONGMAN: NEW YORK

INFO ECON MASS MEDIA POLICY

294 GERSHON, RICHARD A.
THE MAITLAND COMMISSION REPORT: A POLICY REVIEW
1985, MARCH
TELECOMMUNICATION JOURNAL
P83-87

POLICY

295 GILBERG, SHELDON
EDUCATION FOR INDONESIA VIA THE PALAPA SATELLITE
1978, MAY
SATELLITE COMMUNICATIONS
P30-33

INDONESIA SAT EDUC

296 GILLESPIE, T.M.
A TELECOMMUNICATION SYSTEM FOR RURAL VILLAGES
1981, JUNE
TELECOMM JOURNAL
49 N11 P773-6

TELECOMM RURAL ECON SOC
TELEPHONE

297 GILLESPIE, T.M.
TELECOMMUNICATION FOR RURAL VILLAGES
1982, NOVEMBER
TELECOMM JOURNAL
49 N11 P773-6

TELECOMM RURAL SAT

298 GILLING, E.
TELECOMMUNICATION AND ECONOMICS: INTER-COUNTRY COMPARISONS OF
THE CATALYTIC EFFECT OF TELEPHONE SERVICES ON DEVELOPMENT
1975
UNPUBLISHED MBA THESIS

ECON TELEPHONE INTERNAT EVAL
RES WORLD TELECOMM

299 GIMPELSON, LESTER A.
COMMUNICATION PLANNING FOR DEVELOPING COUNTRIES
1974
TELECOMMUNICATION JOURNAL
41 P486-491

PLAN TELECOMM PLAN POLICY

300 GIMPELSON, LESTER A.
PLANNING COMMUNICATIONS SYSTEMS IN DEVELOPING COUNTRIES
1976, JULY
IEEE TRANSACTIONS ON COMMUNICATIONS
COM24 N7 P710

TECH PLAN

301 GIRAUD, A.
LES RESEAUX PENSANTS, TELECOMM ET SOCIETE
1978

TELECOMM SOC EUR FRANCE
CULT

302 GOETSCHIN, P.R.
L'IMPORTANCE DES TELECOMMUNICATIONS
1976, 15 MAY
BULLETIN DE L'ASSOCIATION SUISSE DES ELECTRICIENS

TELECOMM EUR SWITZERLAND TECH
SOC ECON

303 GOLDING, P.
MEDIA ROLE IN NATIONAL DEVELOPMENT: CRITIQUE OF A THEORETICAL ORTHODOXY
1974
JOURNAL OF COMMUNICATION
24 P39-53

MEDIA THEORY RADIO TV
SOC CULT

304 GOLDSCHMIDT, D., FORSYTHE, V., HUN H.
MEDEX GUYANA TWO-WAY RADIO PILOT PROJECT
1980, DEC
ACADEMY FOR EDUCATIONAL DEVELOPMENT: WASHINGTON, D.C.

RURAL GUYANA 2-WAY RADIO EDUC
CARIB HEALTH

305 GOLDSCHMIDT, D., HUDSON, H., AND LYNN W.
TWO-WAY RADIO FOR RURAL HEALTH CARE: AN OVERVIEW
1980, MAY
REPORT/ U.S. AGENCY FOR INTERNAIONAL DEVELOPMENT
ACADEMY FOR EDUCATIONAL DEVELOPMENT: WASHINGTON, D.C.

2-WAY RADIO RURAL HEALTH

306 GOLDSCHMIDT, D., TIETJEN, K., AND SHAW, W.D.
DESIGN AND INSTALLATION OF RURAL TELECOMMUNICATIONS NETWORKS: LESSONS FROM THREE PROJECTS
1986
ACADEMY FOR EDUCATIONAL DEVELOPMENT: WASHINGTON, D.C.

TELECOMM SAT RURAL PLAN
INVEST WORLD BANK POLICY

307 GOLDSCHMIDT, DOUGLAS
FINANCING TELECOMMUNICATION FOR RURAL DEVELOPMENT
1984, SEPT
TELECOMM POLICY (GB)
8 N3 P181-203

FIN TELECOMM RURAL

308 GOLDSCHMIDT, DOUGLAS
TELEPHONE COMM, COLLECTIVE SUPPLY & PUBLIC GOODS:A CASE STUDY OF THE ALASKAN TELEPHONE SYSTEM
1978
UNPUBLISHED DISSERTATION

TELEPHONE ALASKA SOC ECON
N AMER TECHNICAL POLICY

309 GOLDSCHMIDT, DOUGLAS
THE EVOLUTION OF THE ALASKAN SATELLITE TELECOMM SYSTEM
IN TELECOMM & ECONOMIC DEVELOPMENT, POLISHUK & O'BRYANT (EDS)
1977
PAPERS AT INTELCOMM 77

ALASKA SAT TELECOMM HIST
N AMER POLICY TECHNICAL

310 GOLDSCHMIDT, DOUGLAS
THE USE OF THE POST FOR RURAL DEVELOPMENT
1980, MAY 21
PAPER/ ACADEMY FOR EDUCATIONAL DEVELOPMENT, WASHINGTON, D.C.

RURAL

311 GOLDSCHMIDT, DOUGLAS
TWO-WAY RADIO HELPS DELIVER PRIMARY HEALTH CARE
1980 OCT
AID CDC COMM REPORT #32

AF HEALTH AID TELECOMM
2-WAY RADIO RURAL KENYA

312 GOLDSTEIN, IRVING
INTELSAT AND THE DEVELOPING WORLD
1976, JULY
IEEE TRANSACTIONS ON COMMUNICATIONS
COM24 N7 P742-8

INTELSAT TELECOMM

313 GORELIK, M.A. AND EFIMOVA, I.B.
THE ECONOMIC EFFICIENCY OF DEVELOPMENT OF LONG DISTANCE COMMUNICATION
1977
VESTNIK SVIAZI (RUSSIAN)
7

ECON TELEPHONE EUR USSR
RES

314 GORELIK, M.A. EFIMOVA, I.B., AND KARASEVA, E.
HOW TO DETERMINE THE ECONOMIC EFFICIENCY OF LONG DISTANCE TELEPHONE COMMUNICATION
1977
VESTNIK SVIAZI (RUSSIAN)
7

EUR ASIA USSR TELEPHONE
ECON

315 GORELIK, M.A. AND KARASEVA, E.
STANDARDS AND ASSESSMENT OF THE ECONOMIC EFFICIENCY OF LONG DISTANCE TELEPHONE COMMUNICATION
1977
VESTNIK SVIAZI (RUSSIAN)
8

TELEPHONE ECON EUR USSR
RES

316 GOUGH, J. ERIC
POLICY ISSUES IN PLANNING FOR DISTANCE EDUCATION USING A DOMESTIC COMMUNICATION SATELLITE
MARCH 1981
DISTANCE EDUC
2 N1
ERIC # EJ278305

POLICY EDUC AUSTRALIA TV
V-TEXT TELECOMM

317 GOULD, CHARLES L. AND GERBER, C.R.
SKYNET 2000: RAISING GLOBAL PRODUCTIVITY THROUGH SPACE COMMUNICATION
1983 FEB
FUTURIST
17 N1
ERIC # EJ275341

WORLD ECON SAT RELECOMM

318 GOULET, DENIS
THE CRUEL CHOICE: A NEW CONCEPT IN THE THEORY OF DEVELOPMENT
1971
ATHENIUM: NEW YORK

ECON SOC POLICY THEORY

319 GOULET, DENIS
THE UNCERTAIN PROMISE: VALUE CONFLICTS IN TECHNOLOGY TRANSFER
1977
IDOC NORTH AMERICA: NEW YORK

TECH ECON POLIT SOC

320 GOYAL, R.N.
DEVELOPMENT OF TELECOMMUNICATION DURING 1978-83
1980, JUNE
TELECOMM (INDIA)
30 N2 P3-13

TELECOMM TECH ECON INDIA

321 GRAHAM, MICHAEL
SHARING KNOWLEDGE
1983, JANUARY
THE IDRC REPORT
11 N4 P22-24

INFO RES POLICY

322 GRAYSON, L.
EDUCATIONAL SATELLITES: A GOAL . . .?
1976, MAY
IEEE TRANSACTIONS ON EDUC
E-19 N2 P38-45

SAT EDUC CULT SOC

323 GREEN, LYNDSAY AND SIMAILAK, DAVID
INUIT COMMUNICATION NEEDS
1980, MARCH
COMMUNICATION, COMPUTERS AND HUMAN SETTLEMENTS

TELECOMM COMPUTER RURAL
POLICY CANADA

324 GREENBERG, ALLEN
IMPASSE? THE U.S. STAKE IN THIRD WORLD TELECOMMUNICATIONS DEVELOPMENT
1985, SPRING
JOURNAL OF COMMUNICATION
35 N2 P42-49

US TELECOMM

325 GREENBURG, J.S.
EVALUATING THE ECONOMICS OF DESIGN ALTERNATIVES ON DOMESTIC COMMUNICATION SATELLITE ADVENTURES
1979
ACTA ASTRONAUTICA

ECON EVAL SAT TECHNICAL

326 GREGURAS, F.M.
TAX-EXEMPT BOND FINANCING OF COMPUTER, TELECOMM AND OTHER HIGH TECHNOLOGY FACILITIES
1980, FALL
COMPUTER LAW JOURNAL
2 N4 P805-28

FIN COMPUTER TELECOMM TECH
US

327 GRUNDY, B.L., ET AL.
TELEMEDICINE IN CRITICAL CARE: AN EXPERIMENT IN HEALTH CARE DELIVERY
1977, OCTOBER
JOURNAL OF THE AMERICAN COLLEGE OF EMERGENCY PHYSICIANS
6 N10 P439-444

US TELECOMM HEALTH

328 GRUNIG, JAMES E.
INFORMATION DECISION MAKING IN ECONOMIC DEVELOPMENT
1969, AUTUMN
JOURNALISM QUARTERLY
46 N3

INFO ECON

329 GURNSTEIN, MICHAEL
THE SOCIAL IMPACT OF DIGITALIZATION WITH SPECIAL REFERENCE TO DIGITALIZATION IN THE DEVELOPING WORLD
1985, JAN 13-16
PAPER/ PACIFIC TELECOMMUNICATION COUNCIL/ HONOLULU, HAWAII
P392-397

TELECOMM DIGITAL SOC TRAIN
COST

330 GWYN, R.J.
RURAL RADIO IN BOLIVIA: A CASE STUDY
1983, SPRING
JOURNAL OF COMMUNICATION
33 N2 P79-87

BOLIVIA RADIO RURAL

331 HAGELIN, T.M.
PRIOR CONSENT OR THE FREE FLOW OF INFO OVER INTERNAT SAT RADIO & TV:COMPARISON & CRITICISM OF US DOMESTIC & INTERNAT BROADCAST..
1981, MAR
SYRACUSE J OF INTERNAT LAW & COMM
8, N2 265-320

N AMER US SAT RADIO
TV POLICY

332 HAGEN, E.
THE ECONOMICS OF DEVELOPMENT
THE IRWIN SERIES IN ECONOMICS
1975
R.D. IRWIN: HOMEWOOD, ILL.

ECON SOC POLICY THEORY

333 HALL, MADGE
INFORMATION: SPECIALIZED COURSE IN EDUCATION OF THE HEARING IMPAIRED
1983, JUNE
PROJECT/U.W.I. DISTANCE TEACHING EXPERIMENT NETWORK SCHOOL OF ED.

TELECOMM INFO EDUC DOMINICA
JAMAICA ST. LUCIA

334 HALL, MARY D.
TELECOMMUNICATION POLICY FOR THE FUTURE: A MODEL STATE PLAN
1985, JUNE 10
PUBLIC UTILITIES FORTNIGHTLY
115 N1 P15-19

TELECOMM POLICY MODEL PLAN
POLIT

335 HALSAM, C.
HEALTH EDUCATION BY TELEVISION AND RADIO: A CONFERENCE REPORT
1981, MARCH
EDUCATIONAL BROADCASTING INTERNATIONAL
14 N1 P24-29

TV RADIO HEALTH EDUC

336 HAMELINK, CEES
INFORMATION TECHNOLOGY AND THE NEW INFORMATION ORDER
1982, FEB
MEDIA INFORMATION AUSTRALIA
23 P65-71

INFO TECH INTERNAT ECON
POLIT

337 HAMELINK, CEES
INTERNATIONAL FINANCING AND THE INFORMATION INDUSTRY
IN WORLD COMMUNICATION GERBNER, G. & SIEFERT, M. (EDS)
1984

INTERNAT ECON POLIT FIN
INFO

338 HAMELINK, CEES
THE CORPORATE VILLAGE. THE ROLE OF TRANSNATIONAL
CORPORATIONS IN INTERNATIONAL COMMUNICATION
1977

INTERNAT ECON CULT RES
POLIT SOC

339 HAMELINK, CEES
TRANSNATIONAL DATA FLOW IN THE INFORMATION AGE
1984
STUDENTLITTERATUR: LUND

INTERNAT INFO DATA ECON
POLIT POLICY

340 HAMILTON, N.K.
HEALTH EDUCATION BY SATELLITE
1975
HEALTH EDUC MONRAPHS
3 N1 P56-61

EDUC SAT HEALTH EDUC
TELEMEDICINE TECH

341 HANCOCK, A.
COMMUNICATION PLANNING FOR DEVELOPMENT: AN OPERATIONAL FRAMEWORK
1980
MONOGRAPHS ON COMMUNICATION PLANNING, 2
UNESCO: PARIS

THAILAND AFGHANISTAN PHILIPPINES FINLAND
AUSTRALIA PLAN SOC ECON

342 HANCOCK, ALAN
BROADCASTING IN THE THIRD WORLD
1978
EDUCATIONAL BROADCASTING INTERNAT
11, N4

BROADCAST EDUC

343 HANCOCK, ALAN (ED)
TECHNOLOGY TRANSFER AND COMMUNICATION
1984
MONOGRAPHS ON COMMUNICATION PLANNING, 4
UNESCO: PARIS

TELECOMM COMPUTER RURAL POLICY

344 HANNAN, M.
EDUCATION & ECONOMIC DEVELOPMENT: SOME BASELINE MODELS
IN INFORMATION PROCESSES & ECONOMIC DEVELOPMENT, DELACROIX, J. (ED)
1973
UNPUBLISHED PAPER

EDUC ECON THEORY MODEL
INFO

345 HANNAWAY, WYNDHAM
TELE-HEALTH
1981, JULY
TELECOMMUNICATIONS
P52-55

TELECOMM HEALTH

346 HARDY ANDREW P.
THE ROLE OF THE TELEPHONE IN ECONOMIC DEVELOPMENT
1980
TELECOMM POLICY
4 P278-285

TELEPHONE ECON

347 HARDY, ANDREW P.
THE ROLE OF THE TELEPHONE IN ECONOMIC DEVELOPMENT:AN EMPIRICAL
ANALYSIS
1981
ITU: GENEVA

TELEPHONE ECON STAT RES
MODEL

348 HARDY, ANDREW P.
THE ROLE OF THE TELEPHONE IN THE ECONOMY
1980
DISSERTATION ABSTRACTS INTERNATIONAL
40/11-A P5637

TELEPHONE ECON

349 HARLEY, W.G.
THE U.S. SENATE IN THE IPDC
1981, FALL
JOURNAL OF COMMUNICATION
31 N4 P150-158

WORLD RES PLAN

350 HARLOW, MARY MATTHEWS
EXTENDING HEALTH CARE THROUGH TELECOMMUNICATIONS
1975
MITRE MATRIX
8 N2

TELECOMM HEALTH RURAL SAT

351 HARRISON, JAMES
SLOW SCAN TELEVISION IMPROVES HEALTH CARE IN REMOTE COMMUNITIES
1980, SEPTEMBER
COMMUNICATIONS NEWS
P32-33

TV TELECOMM HEALTH REMOTE

352 HARROIS-MONIN F.
CURE DE JOUVENCE ET REMEDES DE CHEVAL POUR UN TELEPHONE CENTENAIRE
1976
SCIENCE ET VIE
701 (FEB)

EUR FRANCE TELEPHONE HIST
CULT SOC

353 HARRY, KEITH ET AL
THE EUROPEAN EXPERIENCE OF THE USE OF MEDIA AND DISTANCE METHODS FOR ADULT BASIC EDUCATION
ED231345
ERIC DOC REPRO SERV

EUR MEDIA TELECOMM EDUC

354 HASEGAWA, H.
CURRENT DEVELOPMENTS & THE FUTURE POLICY OF TELECOMMUNICATION
1983, AUG
ELECTRON INFO & PLAN(INDIA)
10 N11 P689-91

POLICY TELECOMM

355 HAUGHEY, MARGARET & MURPHY, PETER J
PROFESSIONAL CONTINUING EDUCATION FOR RURAL TEACHERS: PROMISING PROSPECTS FOR THE FUTURE
ED238614
ERIC DOC REPRO SERVICE

SAT TELECONF EDUC RURAL
INTERNAT CANADA TELECOMM

356 HAVENS, E.
METHODOLOGICAL ISSUES IN THE STUDY OF DEVELOPMENT
1972
SOCIOLOGIA RURALIS
12

ECON RURAL RES

357 HAVERTY, J.F.
SATNET DEVELOPMENT AND OPERATION PLURIBUS SATELLITE IMPROVES DEVELOPMENT & OPERATION
1983, MAY & FEB
AD-A129 828/0 & 959/7
REP # BBN-5345
BOLT BERANEK & NEWMAN, INC: CAMBRIDGE, MA OR NTIS

SAT TECH MODEL

358 HAWKINS, TONY
TELECOMMUNICATIONS PLANNING AND IMPROVING THE NETWORK
1985
THIRD WORLD DEVELOPMENT
2 P54-55

TELECOMM PLAN ECON

359 HAWKRIDGE, J. AND ROBINSON, J.
ORGANIZING EDUCATIONAL BROADCASTING
1982
UNESCO: PARIS

EDUC BROADCAST

360 HAWORTH, C.
TELECOMMUNICATION POLICY: A SELECTIVE LIST OF PERIODICALS
1984
LIBRARY HIGH TECHNOLOGY
2 N1 P67-72

TELECOMM POLICY

361 HAYDN, R., BONARIUS, H., SCHRAMM, M.
RESEARCH PROJECT MAURETANIA: SATELLITES AS DEVELOPMENT AIDS
1980, FEB
N80-17120/0 OR NASA-TM-76064

SAT MAURETANIA RES

362 HAYWARD, T.E.
ELECTRONIC MARKETING OF AGRIC PRODUCTS IN THE U.S.A. (BASED ON PROCEEDINGS OF THE NATL SYMPOSIUM ON ELECTRONIC MARKETING)
1981, JUNE
AGRIC ABROAD
36 P21-6

AGRIC TECH US

363 HEDEBRO, GOERAN
COMMUNICATION & SOCIAL CHANGE IN DEVELOPING NATIONS. A CRITICAL REVIEW
1982
IOWA STATE UNIVERSITY PRESS: AMES, IOWA

MASS MEDIA SOC RURAL POLIT
L AMER AF ASIA CHINA
CUBA TANZANIA

364 HEDEBRO, GOERAN
COMMUNICATION AND SOCIAL CHANGE IN DEVELOPING COUNTRIES
1979
UNIVERSITY OF STOCKHOLM: SWEDEN

SOC

365 HILL, A.H.
HORSELESS CARRIAGE AND CRACKED MIRROR: THE ELECTRONIC MEDIA IN THE PACIFIC ISLANDS
1982, FEBRUARY
MEDIA INFORMATION AUSTRALIA
N 23 P17-27

PAC POLICY TV MICRONESIA
PAPUA NEW GUINEA POLYNESIA AM SAMOA

366 HILL, A.H. AND MORGAN, M.G.
TELECOMMUNICATION IN ALASKAN VILLAGES
1981, JAN 16
SCIENCE
P241-8

ALASKA TELECOMM RURAL TELEPHONE
TV SAT ECON SOC
CULT REG

367 HILLS, A.H.
TV & TELEPHONE SERVICE IN ALASKAN VILLAGES: A TECHNOLOGICAL, ECONOMIC & ORGANIZATIONAL ANALYSIS
1979
DISSERTATION ABSTRACTS INTERNAT
40/12-B P5759

TV TELEPHONE ALASKA TECH
ECON

368 HITCHENS, H.
LEARNING VIA TELECOMM: READING FROM AUDIOVISUAL INSTRUCTION
1978
ASSOC. FOR EDUC COMM AND TECH

EDUC TELECOMM AV

369 HOBDAY, MICHAEL
THE BRAZILIAN TELECOMMUNICATION INDUSTRY: ACCUMULATION OF
ELECTRONIC TECHNOLOGY IN THE MANUFACTURING AND SERVICE SECTORS
1985, JAN 25
PAPER/U.N. INDUSTRIAL DEVELOPMENT ORGANIZATION (UNIDO)/ NEW YORK

BRAZIL TELECOMM DIGITAL TECH
FIN ECON POLICY

370 HOBDAY, MICHAEL
THE IMPACT OF MICROELECTRONICS ON DEVELOPING COUNTRIES;
THE CASE OF BRAZILIAN TELECOMMUNICATION
1985 APRIL
DEVELOPMENT AND CHANGE, LONDON
N2 P313

COMPUTER TECH TELECOMM BRAZIL
L AMER

371 HOFFMAN, E.
AN ATTEMPT TO EVALUATE REAL BENEFITS OBTAINED DUE TO TELEPHONE
SERVICE UTILIZATION
1974
BIULETYN TECHNICZNY MINISTERSTWA LACZNOSCI (POLISH)
3

TELEPHONE EUR POLAND ECON
SOC EVAL

372 HOFFMAN, E.
EXTENDED EVALUATION OF THE REAL BENEFITS OBTAINED THROUGH
TELEPHONE SERVICE UTILIZATION
1975
BIULETYN TECHNIOZNY MINISTERSTWA LACZNOSCI (POLISH)
2

EUR POLAND TELEPHONE ECON
SOC EVAL

373 HOFFMANN, KURT & RUSH, HOWARD
MICROELECTRONICS, INDUSTRY AND THE THIRD WORLD
1980
FUTURES
12, N4 P289-302

INFO TECH COMPUTER INTERNAT
ECON

374 HOLZBERG, B.J.
THE NEW WORLD INFORMATION ORDER: A LEGAL FRAMEWORK FOR DEBATE
1982, SPRING
JOURNAL OF INTERNAT LAW
14 N2 387-418

LEGAL INTERNAT INFO POLICY

375 HOOVER, BLAINE
A TELECOMMUNICATIONS PLAN FOR THE PHILIPPINES
1983, MARCH 28
TELEPHONY
P46, 47 AND 50

PHILIPPINES TELECOMM

376 HORLEY, A.L.
BROADBAND ELECTRONIC COMMUNICATION FOR BRAZIL: AN INVESTMENT DECISION ANALYSIS
1976
UNPUBLISHED DISSERTATION

BRAZIL TELECOMM INVEST EVAL
L AMER FIN ECON TECH

377 HORNIK, R., SEARLE, B., FOOTE, D., AND MOULTON, J.
THE ROLE OF COMMUNICATIONS IN EDUCATION
1978, DECEMBER
INSTITUTE OF COMMUNICATION RESEARCH: STANFORD UNIVERSITY

TELECOMM EDUC

378 HORNIK, ROBERT
COMMUNICATION AS COMPLEMENT IN DEVELOPMENT
1980
J OF COMM
30 N2 P10-24

TELECOMM RADIO TV SOC
ECON THEORY

379 HORNIK, ROBERT C.
MASS MEDIA USE AND THE "REVOLUTION OF RISING FRUSTRATION":
A RECONSIDERATION OF THE THEORY
1977, OCT
COMMUNICATION RES
4 N4

MASS MEDIA THEORY

380 HORNIK, ROBERT C.
THE IMPACT OF MICROELECTRONICS ON DEVELOPING COUNTRIES: THE
CASE OF BRAZILIAN TELECOMMUNICATIONS
1985
DEVELOPMENT AND CHANGE
16 P313-340

BRAZIL TELECOMM COMPUTER

381 HORNIK, ROBERT ET AL.
COMMUNICATION AS COMPLEMENT IN DEVELOPMENT
1980
JOURNAL OF COMMUNICATION
30 N2 P10-24

TELECOMM RURAL

382 HORNING, JOHN STEPHENS
PUBLIC POLICY TOWARDS RURAL TELEPHONE SERVICE 1950-1981
1982
MIT (DISSERTATION)

POLICY TELEPHONE RURAL

383 HUDSON, H.E. AND PARKER, E.B.
TELECOMMUNICATION: PLANNING FOR RURAL DEVELOPMENT
1975
IEEE TRANSACTIONS ON COMM
23 P1177-1185

TELECOMM PLAN RURAL WORLD

384 HUDSON, H.E.& PARKER, E.B.
MEDICAL COMMUNICATION IN ALASKA BY SATELLITE
1973
NEW ENGLAND J OF MEDICINE
289 (DEC 20)

HEALTH	SAT	ALASKA	RADIO
RES	N AMER		

385 HUDSON, H.E., GOLDSCHMIDT, D.,PARKER, E.B. & HARDY, A.P.
THE ROLE OF TELECOMM IN SOCIO-ECONOMIC DEVELOPMENT: A REVIEW OF THE LITERATURE WITH GUIDELINES FOR FURTHER INVESTIGATION
1981
INTERNATIONAL TELECOMMUNICATION UNION: GENEVA

TELECOMM	SOC	ECON	RES
WORLD			

386 HUDSON, H.E., HARDY, A.P. & PARKER, E.B.
PROJECTIONS OF THE INSTALLATION OF TELEPHONES & THIN ROUTE SATELLITE EARTH STATIONS ON RURAL DEVELOPMENT
1981
INTERNATIONAL TELECOMMUNICATION UNION: GENEVA

TELEPHONE	SAT	RURAL	ECON
RES			

387 HUDSON, HEATHER E.
COMMUNITY COMMUNICATION AND DEVELOPMENT:A CANADIAN CASE STUDY
1974
UNPUBLISHED DISSERTATION

CANADA	SAT	SOC	ECON
TELEPHONE	RURAL	N AMER	

388 HUDSON, HEATHER E.
THE ROLE OF RADIO IN THE CANADIAN NORTH
1977
JOURNAL OF COMMUNICATION
27 N4 P130-139

CANADA	RADIO	SOC	CULT
N AMER			

389 HUDSON, HEATHER E.
THREE CASE STUDIES ON THE BENEFITS OF TELECOMM IN SOCIO-ECONOMIC DEVELOPMENT
1981
INTERNATIONAL TELECOMMUNICATION UNION: GENEVA

TELECOMM SOC ECON RADIO
PAC ALASKA RURAL SAT

390 HUDSON, H.E., HARDY, A.P., AND PARKER, E.B.
PROJECTIONS OF THE IMPACT OF INSTALLATION OF TELEPHONES AND THIN ROUTE SATELLITE EARTH STATIONS ON NATIONAL DEVELOPMENT
1982, DEC
TELECOMMUNICATIONS POLICY
P300-307

TELEPHONE SAT ECON RURAL
AF ASIA LAT AM

391 HUDSON, HEATHER E.
WHEN TELEPHONES REACH THE VILLAGE: THE ROLE OF TELECOMMUNICATIONS IN RURAL DEVELOPMENT
1984
ABLEX: NORWOOD, NJ

RURAL TELEPHONE SOC ECON
AF INDIA PAC ALASKA

392 HUDSON, H.E. ET AL.
THE ROLE OF TELECOMM IN SOCIO-ECONOMIC DEVELOPMENT: A REVIEW OF THE LITERATURE WITH GUIDELINES FOR FURTHER INVESTIGATIONS
1979
INTERNATIONAL TELECOMMUNICATION UNION: GENEVA

TELECOMM SOC ECON SAT

393 HUDSON, H.E., HARDY, A.P. & PARKER, E.B.
IMPACT OF TELEPHONE AND SATELLITE EARTH STATION INSTALLATIONS ON GDP: A MODEL FOR RURAL AREAS OF LDC'S
1982, DEC
TELECOMM POLICY (GB)
6 N4 P300

TELEPHONE SAT RURAL ECON

394 HUDSON, HEATHER E.
A SATELLITE NETWORK IN THE SOUTH PACIFIC
1981, JUNE
UPLINK
P5

SOUTH PAC SAT

395 HUDSON, HEATHER E.
ATS-1 CELEBRATES FIFTEENTH BIRTHDAY A MILESTONE FOR THE PACIFIC
1982, JANUARY
SATELLITE COMMUNICATIONS
6 N1 P22-23

PAC SAT

396 HUDSON, HEATHER E.
SATELLITE COMMUNICATIONS AND DEVELOPMENT: FROM CONJECTURE TO REALITY
1985
SPACE COMMUNICATION AND BROADCASTING
3 N4 P289-295

SAT EDUC

397 HUDSON, HEATHER E.
DEMAND AND NEED: PROBLEMS IN PLANNING RURAL TELECOMMUNICATIONS
1985, FALL
TELEMATICS AND INFORMATICS
2 N3 P251-258

TELECOMM RURAL REMOTE PLAN
POLICY REG FIN

398 HUDSON, HEATHER E.
IMPLICATIONS FOR DEVELOPMENT COMMUNICATIONS: THE U.S. FACES WARC
1979, WINTER
J OF COMMUNICATION
29 N1 P126-186

SAT INFO TECH RURAL
WARC

399 HUDSON, HEATHER E.
MODERN TECHNOLOGIES CAN SERVE TRADITIONAL CULTURES
1983, MARCH
INTERMEDIA
11 N2 P39-48

TELECOMM TECH CULT

400 HUDSON, HEATHER E.
NEW COMMUNICATION TECHNOLOGIES: POLICY ISSUES FOR THE DEVELOPING WORLD
1986
INTERNATIONAL POLITICAL SCIENCE REVIEW
7 N6 P332-343

TELECOMM POLICY INTERNAT SAT

401 HUDSON, HEATHER E.
SATELLITE COMMUNICATION AND DEVELOPMENT: A REASSESSMENT
ED236706
ERIC DOC REPRO SERV

SAT TELECOMM SOC ECON
RURAL

402 HUDSON, HEATHER E.
SATELLITE COMMUNICATIONS IN THE PACIFIC ISLANDS: STATUS AND STRATEGIES
1979
ACADEMY FOR EDUCATIONAL DEVELOPMENT: WASHINGTON, D.C.

PAC SAT TELECOMM

403 HUDSON, HEATHER E.
SATELLITES IN PACIFIC DEVELOPMENT: THE NEED FOR TRANSITION
1979
PROCEEDINGS OF THE PACIFIC TELECOMM CONFERENCE: HONOLULU, HAWAII

PAC SAT SOC

404 HUDSON, H.E., FORSYTHE, V., AND BURNS, S.
KEEPING IN TOUCH BY TWO-WAY RADIO
1983
WORLD HEALTH FORUM
V4 P157-161

HEALTH 2-WAY

405 HUDSON, HEATHER E.
THE ROLE OF TELECOMMUNICATION IN THE DEVELOPMENT PROCESS: RURAL TELECOMMUNICATION IN DEVELOPING COUNTRIES
PAPER/ PACIFIC TELECOMM PROCEEDINGS
HONOLULU, HAWAII

TELECOMM RURAL

406 HUDSON, HEATHER E. AND PARKER, EDWIN B.
TELECOMMUNICATION PLANNING FOR RURAL DEVELOPMENT
1975, OCT
IEEE TRANSACTIONS ON COMMUNICATIONS
23 N10 1177-1185

RURAL PLAN SAT TELEPHONE

407 HUDSON, HEATHER E.
SATELLITES AND THE DEVELOPING WORLD: THE 1985 SPACE WARC
1985, DEC
TRANSNATIONAL DATA REPORT

WARC

408 HUPE, HOWARD H.
COMING BROADCAST SATELLITES -- WHERE THEY SHOULD AND SHOULD NOT BE USED IN DELIVERING EDUCATIONAL SERVICES
1974
EDUCATIONAL TECHNOLOGY
14 N12 P40-3

EDUC SAT TECH

409 HUPE, HOWARD H.
EDUCATIONAL SATELLITES - COST AND EFFECTS ON EDUCATIONAL SYSTEM
1974
EDUCATIONAL TECHNOLOGY
14 N10 P48-52

EDUC SAT TECH COST
SOC

410 HURD, JANE
SATELLITE COMMUNICATIONS FOR THE PACIFIC ISLANDS: AN ALTERNATIVE TO ISOLATION
1983
PACIFIC ISLANDS COMMUNICATION JOURNAL
12 N2

PAC SAT

411 HURLEY, N.P.
UNIVERSITY SATELLITE FOR LATIN AMERICA
1975
J OF COMMUNICATION
25 N2 P157-64

EDUC SAT L AMER

412 HUSARSKI, L.
EFFICIENCY OF LONG DISTANCE TELEPHONE SERVICE IN THE SOVIET NATIONAL ECONOMY
1976
WIADOMOSCI TELEKOMUNIKACYJNE
6

USSR TELEPHONE EUR ECON
RES

413 HUSY, W. JOHN
ECONOMIC BENEFITS OF OPERATIONAL ENVIRONMENTAL SATELLITES
1983, MAR
REP NOAA-83082409

SAT ECON DATA

414 HUSY, W. JOHN & HEACOCK, E. LARRY
THE ECONOMIC BENEFITS OF ENVIRONMENTAL SATELLITES
1978, APRIL
PB-292 340/7 OR REPT NOAA-79012303

SAT ECON DATA EVAL

415 IDB NEWS
TELECOMMUNICATION IN LATIN AMERICA: STILL MORE TO BE BUILT
1977
INTER-AMER DEVELOPMENT BANK
4 (7)
INTER-AMER DEVELOPMENT BANK WASH D.C.

TELECOMM L AMER PLAN POLICY

416 IDRC
COMPUTER-BASED CONFERENCE SYSTEMS FOR DEVELOPING COUNTRIES:
REPORT OF A WORKSHOP HELD IN OTTAWA, CANADA, 26-30 OCT. 1981
1982, 3

COMPUTER TELECONF

417 INDIAN POSTS & TELECOMM BOARD
SURVEY OF PUBLIC CALL OFFICES
1978

ASIA INDIA TELEPHONE RURAL
STAT RES

418 INDIAN POSTS AND TELECOMMUNICATIONS BOARD
SURVEY OF PUBLIC CALL OFFICES
1982
NATIONAL COUNCIL OF APPLIED ECONOMIC RESEARCH: AMHERST, MA

INDIA TELEPHONE SURVEY

419 INGLE, HENRY T.
COMMUNICATION MEDIA AND TECHNOLOGY: A LOOK AT THEIR ROLE
IN NON-FORMAL EDUCATION PROGRAMS
1974
INFO BULLETIN
N5
THE INFO CTR ON INSTRUCTIONAL TECH, ACADEMY FOR ED DEV

EDUC EVAL

420 INKELES, A.
MAKING MEN MODERN: ON THE CAUSES AND CONSEQUENCES OF CHANGE
IN SIX DEVELOPING COUNTRIES
1969
AMER J OF SOCIOLOGY
75 P208-225

SOC CULT EDUC THEORY

421 INNIS, HAROLD
BIAS OF COMMUNICATION
1951
UNIVERSITY OF TORONTO PRESS: TORONTO

N AMER CANADA TELECOMM ECON
CULT THEORY

422 INNIS, HAROLD
EMPIRE AND COMMUNICATION
1950
CLARENDON PRESS: OXFORD

N AMER CANADA TELECOMM ECON
CULT THEORY

423 INSTITUTE FOR COMMUNICATION RESEARCH, STANFORD UNIV
THE ROLE OF THE TELEPHONE IN ECONOMY
1980
INFO GATEKEEPERS

TELEPHONE ECON RES

424 INTERNATIONAL BANK FOR RECONSTRUCTION AND DEVELOPMENT
ECONOMIC EVALUATION OF PUBLIC UTILITY DEPT
1974

WORLD BANK EVAL ECON

425 INTERNATIONAL TELECOMMUNICATION UNION
APPROPRIATE MODERN TELECOMMUNICATION TECHNOLOGY FOR INTEGRATED RURAL DEVELOPMENT IN AFRICA
1981, NOVEMBER
ITU: GENEVA

AF TELECOMM TECH RURAL
ITU

426 INTERNATIONAL TELECOMMUNICATION UNION
REPORTS ON THE PEACEFUL USES OF OUTER SPACE
SINCE 1962
ITU: GENEVA

ITU TECH UN WORLD

427 INTERNATIONAL TELECOMMUNICATION UNION
RURAL TELECOMMUNICATION
1981, NOVEMBER
ITU: GENEVA

TELECOMM RURAL ITU

428 INTERNATIONAL TELECOMMUNICATION UNION
SCIENCE, TECHNOLOGY & SOCIETY. NEEDS, CHALLENGES AND LIMITATIONS. EDS. K.H. STANDKE, M. ANANDARKRISHNAN
1979
207-238
PERGAMON PRESS: NY

TELECOMM INTERNAT ECON

429 INTERNATIONAL TELECOMMUNICATION UNION
SEMINAR ON THE PLANNING AND DEVELOPMENT OF TELECOMM NETWORKS
OUTSIDE OF LARGE CITIES AND THE MAINTENANCE OF TELECOMM SERV
1972
ITU: GENEVA

ITU PLAN TELECOMM RURAL

430 INTERNATIONAL TELECOMMUNICATION UNION
TELECOMMUNICATION AND DEVELOPMENT
1978
BOOKLET #22
INTERNATIONAL TELECOMMUNICATION UNION: GENEVA

WORLD ITU TELECOMM

431 INTERNATIONAL TELECOMMUNICATION UNION
YEARBOOK OF COMMON CARRIER TELECOMM STATISTICS AND RADIO
COMMUNICATION STATISTICS
ANNUAL
ITU: GENEVA

STAT TELECOMM RADIO TELEPHONE

432 INTERNATIONAL TELECOMMUNICATION UNION
GLODOM: CHEAP TELEPHONE CALLS FOR THE THIRD WORLD
1985, JAN
INTERMEDIA

WORLD RADIO TELEPHONE SAT
RURAL FIN ITU COST
REMOTE

433 INTERNATIONAL TELEGRAPH & TELEPHONE CONSULTATIVE COMMITTEE
ECONOMIC STUDIES AT THE NATIONAL LEVEL IN THE FIELD OF
TELECOMM (GAS 5 MANUALS)
1968,'72,'76
ITU: GENEVA

ECON RES TELECOMM

434 IZZO, L.L.C.
RURAL TELEPHONY IN THE STATE OF SAO PAOLO, BRAZIL.
IN POLISHUK & O'BRYANT, EDS, TELECOMM & ECON DEVELOPMENT
1977
HORIZON HOUSE INTL: DEDHAM, MA

BRAZIL TELEPHONE RURAL L AMER

435 JACKSON, KARL D. & MOELIONO, JOHANNES
COMMUNICATION AND NATIONAL INTEGRATION IN SUDANESE VILLAGES:
IMPLICATIONS FOR COMMUNICATION STRATEGY
1972
EWCI PAPER
N3
EAST-WEST COMM INST: HONOLULU, HAWAII

M EAST SUDAN PLAN

436 JAHR, DALE
PUBLIC POLICY CONSIDERATIONS OF PRICING TELEPHONE SERVICES:
A STAFF STUDY
1984
GPO IEM 1000-B, 1000-C (MICROFICHE)

POLICY TELEPHONE COST RES
POLIT US

437 JAMISON, D.T. AND MCANANY, E.G.
RADIO FOR EDUCATION AND DEVELOPMENT
1978
SAGE PUBLISHING: BEVERLY HILLS, CA.

RADIO EDUC

438 JAMISON, D.T., KLEES, S., AND WELLS, S.J.
THE COSTS OF EDUCATIONAL MEDIA: GUIDELINES FOR PLANNING
AND EVALUATION
1978
SAGE: BEVERLY HILLS, CA

EDUC COST

439 JAMISON, DEAN T.
BASIC EDUCATION AND EXTENSION COSTS, EFFECTS, AND ALTERNATIVES
1982
WORLD BANK: WASHINGTON, D.C.

EDUC COST

440 JANKY, J. AND BAREWALD, J.
INTERFERENCE CONTROL IN BROADCAST SATELLITE APPLICATIONS: ANTENNA SIDELOBE PATTERNS AND TRANSPONDER GAIN
1977
EDUTEL COMMUNICATIONS AND DEVELOPMENT: PALO ALTO, CA

SAT TECH

441 JANNES H.
LE DOSSIER SECRET DU TELEPHONE
1970
PARIS

TELEPHONE

442 JASENTUL, N.,CORREA, L., SINGER, I., NICHOLS, J.C., CHAYES, A.
DIRECT SATELLITE BROADCASTING AND THIRD WORLD
1974
COLUMBIA JOURNALISM QR
13 N1 P68-81

SAT BROADCAST TV ECON

443 JASSAL, T.P.S.
THE SOCIETIES OF TELECOMMUNICATION
1981, JUNE
TELECOMM (INDIA)
31 N1 P3-7

SOC ECON TELECOMM INDIA

444 JAYAWEERA, NEVILLE D.
NEW COMMUNICATION TECHNOLOGIES AND THEIR IMPACT ON WESTERN INDUSTRIALIZED COUNTRIES
1984
108-127
FRIEDRICH EBERT STIFTUNG: BONN

TECH SAT TELECOMM RURAL
SOC POLIT POLICY

445 JEANCHARLES, N.J., POPOVICS, L., ROYER, J., & VERDIER, J.
EXAMEN DE QUELQUES SOLUTIONS TECHNIQUES AU SERVICE DES TELECOMMUNICATIONS RURALES
1981
CIT ALCATEL: PARIS, FRANCE

TELECOMM RURAL

446 JENKINS, THOMAS M.
HOSPITALS PRESCRIBE TELECONFERENCES TO CURE HEALTH CARE EDUCATION ILLS
1983, AUG 1
TELEPHONY
P26-29

TELECONF HEALTH EDUC INFO

447 JENNINGS, R.D.
THIN-ROUTE RURAL COMMUNICATION USING INTELSAT SATELLITE
1984
COMM SAT SYSTEMS CONF 10TH/ ORLANDO FLORIDA TECHNICAL PAPERS
A84-2525110-32
AMER INST OF AERONAUTICS AND ASTRONAUTICS

RURAL SAT INTELSAT TECHNICAL
WORLD

448 JENNY, F.F.
DIGITAL SOLUTION: KEY TO GROWTH FOR RURAL AREAS (TELECOMM)
1980, DEC
TELEPHONE ENG AND MGT
84 N24 P63-6

TELECOMM RURAL DIGITAL

449 JEQUIER, NICOLAS
TELECOMMUNICATION FOR DEVELOPMENT: FINDINGS OF THE
ITU-OECD PROJECT
1984, JUNE
TELECOMM POLICY (GB)
8 N2 P838

TELECOMM ITU RES

450 JEQUIER, NICOLAS
THE INDIRECT EMPLOYMENT-GENERATIVE EFFECTS OF INVESTMENTS IN
TELECOMMUNICATION
1983
PAPER AT WORLD COMM YEAR SEMINARS ON TELECOMM FOR DEVELOPMENT

ECON INVEST TELECOMM

451 JESSOP, C.W.A., YEOH, B.M., PARRY, C.W.
ECONOMIC ANALYSIS-UNCERTAINTY & RISK III (TELECOMM)
1983
TELECOMM J AUST
33 N3 227-30

ECON EVAL TELECOMM

452 JET PROPULSION LAB
SATELLITE DATA DISTRIBUTION SYSTEM. DEVELOPMENT & DEMONSTRATION
OF NEW FISHERIES SUPPORT PRODUCTS BASED ON REMOTE OCEAN SENSING
1980, MAY
N83-32147/1
NAS 1.15:85334
NASA: WASHINGTON, D.C.

SAT DATA MODEL NASA

453 JOHNSON, J.L.
TELCOS MUST CHART A CHANGING COURSE (TELECOMM
SERVICES)
1983, NOV 21
TELEPHONY
205, N22

TELEPHONE BUSI EVAL PLAN

454 JONSCHER, CHARLES
ASSESSING THE BENEFITS OF TELECOMMUNICATIONS
1985, JANUARY
INTERMEDIA
13 N 1 P21-24

TELECOMM ECON

455 JONSCHER, CHARLES
ECONOMIC CAUSES OF THE GROWTH IN THE INFORMATION SECTOR
1982, MAY
INFORMATION ECONOMICS AND POLICY
1 N1

TECH ECON INFO

456 JONSCHER, CHARLES
THE CONTRIBUTION OF THE COMMUNICATIONS SECTOR TO AFRICAN DEVELOPMENT
1985
PAPER/ECONOMIC COMMISSION FOR AFRICA
P1-19

AF ECON

457 JONSCHER, CHARLES
THE ECONOMIC ROLE OF TELECOMMUNICATIONS
IN MOSS, M. ED.; TELECOMMUNICATIONS AND PRODUCTIVITY
1981

TELECOMM ECON

458 JONSCHER, CHARLES
THE ROLE OF ECONOMIC DEVELOPMENT
1984, APRIL
LONDON, ENGLAND

ECON

459 JUSSAWALLA, M. AND CHEA, C. (EDS.)
INTERNATIONAL TRADE AND INFORMATION ECONOMICS
1984
COMMUNICATION AND INFORMATION ECONOMICS
5 P52-69
ELSEVIER/NORTH HOLLAND: AMSTERDAM

INTERNAT SOC ECON THEORY
DATA INFO

460 JUSSAWALLA, M. AND CHEAH, C.W.
TOWARDS AN INFORMATION ECONOMY: THE CASE OF SINGAPORE
1983
INFORMATION ECONOMICS AND POLICY
1 N2 P161-176

SINGAPORE INFO ECON

461 JUSSAWALLA, M. AND LAMBERTON, D.M., EDS.
COMMUNICATION ECONOMICS AND DEVELOPMENT
(PERGAMON POLICY STUDIES ON INTERNATIONAL DEVELOPMENT)
1982
PERGAMON PRESS: NEW YORK

JAPAN INTERNAT TELECOMM ECON

462 JUSSAWALLA, MEHEROO
COMMUNICATION AND THE GLOBAL ECONOMIC ORDER
1978, JULY
DEVELOPMENT COMMUNICATION REPORT
N3

ECON WORLD

463 JUSSAWALLA, MEHEROO
COMMUNICATION TECHNOLOGY AND INTERNATIONAL BANKING: THEORY VS
REALITY (INSTITUTE OF CULTURE AND COMMUNICATION, EAST-WEST CENTER)
1985, MARCH
PAPER/ 19TH INTERNATIONAL ATLANTIC ECONOMIC CONFERENCE, ROME

FIN TECH

464 JUSSAWALLA, MEHEROO AND D.M. LAMBERTON (EDS)
COMMUNICATION ECONOMICS AND DEVELOPMENT
1982
PERGAMON PRESS: ELMSFORD, NY

ECON PLAN POLICY RES

465 JUSSAWALLA, MEHEROO
POLICY OUTCOMES OF WARC '79: THEIR IMPACT ON DEVELOPING COUNTRIES
1981
MEDIA ASIA
8 N1 P7-12

WARC POLICY ECON MGT
AGRIC SPECTRUM SAT

466 JUSSAWALLA, MEHEROO
THE ECONOMIC IMPLICATIONS OF SATELLITE TECHNOLOGY AND THE INDUSTRIALIZATION OF SPACE
1984, SEPT
TELECOMMUNICATIONS POLICY
8 N3 P237-248

ORBIT SAT TECH ECON

467 JUSSAWALLA, MEHEROO
THE ECONOMICS OF TELECOMMUNICATION FOR DEVELOPMENT
(EAST-WEST COMMUNICATION CENTER)
P1-11
EAST-WEST COMMUNICATION CENTER: HONOLULU, HAWAII

INDIA TELECOMM ECON RES

468 KAMAD, A., DESSONKI, A., AND POOL, I.
COMMUNICATION SYSTEM IN RURAL EGYPT
1980
REPORT/COMMUNITY NEEDS FOR RURAL DEVELOPMENT RESEARCH PROJECT
CAIRO UNIVERSITY, MIT

EGYPT RURAL RES

469 KAMAL, A.A.
A COST BENEFIT ANALYSIS OF RURAL TELEPHONE SERVICE IN EGYPT
1981
INTERNATIONAL TELECOMMUNICATION UNION: GENEVA

COST RURAL TELEPHONE EVAL
EGYPT

470 KAMMAN, ALAN B.
COMMUNICATION AND DEVELOPMENT: THE THIRD WORLD PRIORITY AND THE U.S. INTEREST
1979, DEC.
ARTHUR D. LITTLE: LEXINGTON, MA

WORLD US

471 KAMMAN, ALAN B.
HIGHER PRIORITY FOR COMMUNICATIONS INFRASTRUCTURE DEVELOPMENT
1983
TELECOMMUNICATIONS FORUM PROCEEDINGS
INTERNATIONAL TELECOMMUNICATION UNION: GENEVA

WORLD TELECOMM STAT ITU
RURAL

472 KAMMAN, ALAN B.
THE AVAILABILITY OF CAPITAL FOR TELECOMMUNICATION PROJECTS
1985
TELEMATICS AND INFORMATICS
2 N1 P55-59

TELECOMM FIN

473 KANGA, R.
INFORMATION TECHNOLOGY
THIRD JERUSALEM CONFER ON INFO TECH ED. JOSEF MONETA
1978

COMPUTER INTERNAT EVAL RES
ECON SOC

474 KAPLAN, DON
THE 40,000-MILE PHONE CALL
1984, FEB
LEARNING
12 N7

SAT EDUC TELECOMM

475 KARAMCHANDANI, LAL
TELEVISION FOR RURAL DEVELOPMENT: INDIAN EXPERIENCE WITH SITE
1976
TRAINING FOR AGRICULTURE AND RURAL DEVELOPMENT
P 127-135

INDIA SAT TV AGRIC
REMOTE

476 KARUNARATNE, N.E.
TELECOMMUNICATION & INFORMATION IN DEVELOPMENT PLANNING STRATEGY
IN M. JUSSAWALLA & D.M. LAMBERTON (EDS) COMM ECONOMICS & DEVELOPMENT
1982

TELECOMM PLAN ECON

477 KASSNER, M.H.
PROFILES OF TELECOMMUNICATION DEVELOPMENT; THE DISTRIBUTION OF
TELEPHONE DENSITY IN TERMS OF URBAN OR RURAL DEVELOPMENT
1978, JUNE
ASDA LIMITED COMMUNICATION: ONTARIO, CANADA

TELECOMM TELEPHONE URBAN RURAL
ECON

478 KATZ, E., AND WEDELL, G.
BROADCASTING IN THE THIRD WORLD: PROMISE AND PERFORMANCE
1977
HARVARD UNIVERSITY PRESS: CAMBRIDGE, MA

BROADCAST RADIO ALGERIA BRAZIL
CYPRUS INDONESIA IRAN NIGERIA
PERU SENEGAL

479 KATZMAN, NATAN
THE IMPACT OF COMMUNICATION TECHNOLOGY: SOME THEORETICAL PREMISES AND THEIR IMPLICATION
1974, AUGUST
EKISTICS 225

TECH TEHORY

480 KAUL, S.N.
BENEFITS OF RURAL TELECOMM IN DEVELOPING COUNTRIES
1979, DEC
TELECOMM (INDIA)
29 N2

TELECOMM RURAL

481 KAUL, S.N.
INDIA'S RURAL TELEPHONE NETWORK
ECONOMICS STUDY CELL POSTS AND TELEGRAPH BOARDS
1981
MINISTRY OF COMMUNICATIONS OF INDIA: NEW DELHI

INDIA TELEPHONE RURAL

482 KAUL, S.N.
THE INTEGRATION OF TELECOMM SERVICES PLAN WITH PLANNING IN OTHER SECTORS
1978
CANADIAN DEPT OF COMMUNICATIONS: OTTAWA

TELECOMM PLAN

483 KEINSTEAD, PHILLIP
SATELLITE BRINGING RENAISSANCE TO RADIO NETWORKING
1981, OCTOBER
BROADCAST COMMUNICATION
P58-69

SAT RADIO

484 KENNEY, G.I., GUITE, M. AND HUDSON, H.E.
TWO-WAY RADIO FOR RURAL HEALTH DELIVERY IN LESOTHO
1980, JANUARY
PAPER/ ACADEMY FOR EDUC DEVELOPMENT, WASHINGTON, D.C.

LESOTHO 2-WAY RADIO RURAL HEALTH

485 KENNY, D.A.
CROSS-LAGGED PANEL CORRELATION: A TEST FOR SPURIOUSNESS
1975
PSYCHOLOGICAL BULLETIN
82 P887-93

STATS

486 KERR, W.T. AND BLEVIS, B.C.
TELECOMMUNICATION SERVICES FOR RURAL AND REMOTE AREAS
1984
TELEMATICS AND INFORMATICS
1 N1 P37-46

CANADA TELECOMM RURAL REMOTE
HEALTH EDUC

487 KEUNE, REINHARD
TV NEWS IN A NORTH-SOUTH PERSPECTIVE. REPORTS-DOCUMENTS-
RECOMMENDATIONS OF THE INTERNATIONAL BROADCAST NEWS WORKSHOP
1981 FEB23-25

BROADCAST POLIT

488 KHATTAB, M. K.
IMPORTANCE OF TELECOMMUNICATION MEDIA USED IN CONJUNCTION W/
NON-FORMAL EDUC & RURAL COMM CTRS FOR RURAL DEVELOPMENT SAUDI ARABIA
1982
DISS. ABST. INTERNAT
43/10-A P3148

M EAST SAUDI ARABIA TELECOMM MEDIA
EDUC

489 KILGOUR, FREDERICK G.
PUBLIC POLICY AND NATIONAL AND INTERNAT NETWORKS
SEP 1983
INFO TECH AND LIBRARIES
2 N3 P239-45

INTERNAT EUR WORLD POLICY
TELECOMM INFO

490 KILGOUR, MARY CAMERON
THE TELEPHONE IN THE ORGANIZATION OF SPACE FOR DEVELOPMENT
1982
DISSERTATION ABSTRACTS INTERNAT
43/05-A P1686

TELEPHONE

491 KINI, M.M.
PROBLEMS OF PLANNING RURAL TELECOMM DEVELOPMENT IN THE
DEVELOPING COUNTRIES OF THE ESCAP REGION
1978,SEPT 11-22
SEMINAR/ RURAL TELECOMMUNICATIONS
NEW DELHI, INDIA

PLAN RURAL TELECOMM

492 KINN, ROBERT
UNITED STATES PARTICIPATION IN THE INTERNATIONAL TELECOMMUNICATION
UNION AND SERIES OF INTERVIEWS
1985, WINTER
THE FLETCHER FORUM
9 N1 P37-68

US ITU

493 KITAHARA, Y.
RECENT DEVELOPMENTS AND THE FUTURE POLICY OF TELECOMMUNICATION
1983, JAN
JAPANESE TELECOMM REVIEW
25 N1 P2-6

JAPAN TELECOMM POLICY TECH

494 KLAUS, JONATHAN G.
RURAL SATELLITE COMMUNICATIONS SYSTEM NETWORK CONSIDERATIONS
1982
PAPER/ GLOBECOM `82
MIAMI

RURAL SAT

495 KLEES, S.J. AND WELLS, S. J.
COST ANALYSIS FOR EDUCATIONAL DECISION MAKING
1978, NOVEMBER
#AID/TA-C-1348

EDUC COST

496 KLEES, STEVEN J.
TELEVISION AS AN EDUCATIONAL MEDIA: THE CASE OF MEXICAN SECONDARY EDUCATION
1979
COMPARATIVE EDUCATIONAL REVIEW

MEXICO EDUC TV

497 KLING, ROBERT
VALUE CONFLICTS IN COMPUTING DEVELOPMENTS: DEVELOPED AND DEVELOPING COUNTRIES
1983, MARCH
TELECOMMUNICATIONS POLICY
7 N1 P12-34

US BRAZIL COMPUTER SOC
POLIT

498 KOCHEN, M. & DEUSH, K.W.
TOWARD A RATIONAL THEORY OF DECENTRALIZATION:SOME IMPLICATIONS OF A MATHEMATICAL APPROACH
1969
AMERICAN POLIT SCIENCE REVIEW
63 P734-49

STATS THEORY

499 KOELLE, D.F.
THE DEVELOPMENT OF GEOSTATIONARY COMMUNICATION SATELLITES
1981, JULY
N82-2586/7

SAT ORBIT SPECT

500 KOKUSAI, KYORYOKU JIGYODAN
THE REPUBLIC OF INDONESIA: FEASIBILITY STUDY REPORT ON SULAWESI TERRITORIAL NETWORK PROJECT
1983, FEBRUARY

INDONESIA

501 KOUROURMA, M.Y.
CONSIDERATIONS SOCIALES SUR LA PLANIFICATION DU DEVELOPPEMENT DES TELECOMM DANS LES REGIONS ISOLEES ....
1978
CANADIAN DEPARTMENT OF COMMUNICATIONS: OTTAWA

AF UPPER VOLTA RURAL
TELECOMM PLAN SOC

502 KREIMER, O.
TELECOMMUNICATIONS IN MEDICINE: INTERACTIVE SATELLITE RADIO FOR HEALTH CARE IN VILLAGE ALASKA
1975
STANFORD UNIV. INSTITUTE FOR COMM RESEARCH: STANFORD, CA.

TELECOMM ALSAKA RURAL SAT
RADIO REMOTE

503 KRIEGL, W., LAUFENBERG, W.
COMMUNICATIONS SATELLITE SYSTEM FOR AFRICA
1980, SEPT
N81-24160/6

AF SAT

504 KUES, J.R.
PROVIDING INFO TO HEALTH-CARE PROFESSIONALS VIA TELECOMMUNICATION - A CASE STUDY
1980
PROCEEDINGS OF THE AMERICAN SOCIETY FOR INFO SCIENCE
17 P112-4

INFO HEALTH SOC TELECOMM
RURAL

505 KURLAND, J.P., SAWITZ, AND OSBORNE, G.
ASSESSMENT OF CANDIDATE COMMUNICATIONS SATELLITE FOR DEVELOPING COUNTRIES
1977
REPORT/U.S. AGENCY FOR INTERNATIONAL DEVELOPMENT/SILVER SPRING, MD.

TELECOMM SAT EVAL

506 KWELI, K. RANDOLPH, M.L.
SATELLITE TELECONFERENCE - A NEW TOOL FOR COMMUNITY AND ECONOMIC DEVELOPMENT
1984, WINTER
URBAN LEAGUE REVIEW
8 N1 P40-54

SAT TELECONF ECON

507 LAFFAY, J.
LES TELECOMMS (QUE SAIS-JE?,335)
1968

TELECOMM

508 LAFLIN, MICHAEL (ED.)
THE UNIVERSITY OF THE SOUTH PACIFIC: A COURSE OF ACTION
1985, NOVEMBER
REPORT/INTERNATIONAL RESEARCH ACADEMY FOR EDUCATIONAL DEVLOPMENT

PAC TELECONF AID

509 LAFON, M.
LE FINANCEMENT DES TELECOMMS
1978
REVUE FRANCAISE DES TELECOMMS, PARIS
N 27 P 40

TELECOMM

510 LALOR, GERALD C.
CARIBBEAN REGIONAL COMMUNICATIONS SERVICE STUDY
REPORT
ED240985
ERIC DOC REPRO SERV

CARIB TELECOMM EDUC

511 LALOR, GERALD C.
PROJECT SATELLITE REPORT
1978, JUNE
UNIVERSITY OF WEST INDIES: MONA, KINGSTON, JAMAICA

SAT

512 LANCASTER, K.L. (ED.)
INTERNATIONAL TELECOMMUNICATIONS
1982
ARTHUR D. LITTLE: LEXINGTON, MA

INTERNAT TELECOMM BUSI COMPUTER
JAPAN FIBER SAT

513 LAND, THOMAS
MICROWAVE TECHNOLOGY HIGHLIGHTS DOMESTIC TELECOMM INADEQUACIES
1984, OCTOBER
AFRICAN BUSINESS
P88

TELECOMM AF M-WAVE

514 LANGE, JAMES C.
EDUCATIONAL AND CULTURAL SATELLITE EXCHANGE IN MICRONESIA
1983
PACIFIC COMMUNICATION JOURNAL
12 N2 P130-146

MICRONESIA SAT EDUC CULT

515 LANGE, JAMES ET AL.
ATS-1 SOCIAL SERVICE SATELLITE NETWORK IN THE PACIFIC
1984, JAN 8-11
PROCEEDINGS PACIFIC TELECOMMS CONFERENCE: HONOLULU, HI
P109-118

PAC SAT SOC

516 LANNON, LARRY
TELECOMMUNICATION IN MEXICO: AMBITIOUS PLANS DESPITE SERIOUS FINANCIAL PROBLEMS
1985, APRIL 22
TELEPHONY
208 N16 P28-32

TELECOMM MEXICO PLAN FIN
EVAL

517 LAUFFER, S. AND CASEY-STAHMER, A.
TELECOMMUNICATION SYSTEMS FOR EDUCATION AND TRAINING
1983
ACADEMY FOR EDUCATIONAL DEVELOPMENT: WASHINGTON, D.C.

TELECOMM EDUC TRAIN HEALTH
AGRIC TELECONF TV 2-WAY RADIO
TELEX COMPUTER

518 LAUFFER, S.
REVIEW AND ANALYSIS OF RURAL TELECOMMUNICATION IN DEVELOPING COUNTRIES
1984, NOV 30
ACADEMY FOR EDUCATIONAL DEVELOPMENT: WASHINGTON, D.C.

TELECOMM TELEPHONE RURAL EVAL

519 LAZER, E.A., ELTON, M.C.J., AND JOHNSON, J.W.
THE TELECONFERENCING HANDBOOK: A GUIDE TO COST EFFECTIVE COMMUNICATION
1983
KNOWLEDGE INDUSTRY PUBLICATIONS, INC: WHITE PLAINS, NEW YORK

TELECONF FIN COST

520 LEBEAU, A.
THE DEVELOPMENT OF SATELLITE COMMUNICATION AND ITS SOCIOECONOMIC IMPLICATIONS
1982
EARTH-ORIENTED APPLICATIONS OF SPACE TECH
2 N3-4 221-4

SAT HIST TECH SOC
ECON

521 LEE, H.R.
SATELLITE COMMUNICATIONS IN THE PUBLIC SERVICE
1977
MEDIA ASIA
4 N314 P63-67

PAC SAT SOC CULT
INFO EDUC HEALTH

522 LEE, R.
THE FLOW OF INFORMATION TO DISADVANTAGED FARMERS
1982
UNPUBLISHED DISSTERTATION

RURAL AGRIC INFO

523 LEET, GLEN
HOW TO GAIN ACCESS TO INFORMATION ON SCIENCE AND TECHNOLOGY FOR DEVELOPMENT FROM DATA BASES VIA COMPUTER TELECOMMUNICATION
1979, AUG
PB80-133143

COMPUTER INFO TECH TELECOMM
DATA

524 LEET, GLEN
THE USE OF MODERN COMPUTER TELECOMM SYSTEMS TO ENCOURAGE
THE PARTICIPATION OF PEOPLE IN DEVELOPMENT
1979, APRIL 21
PB80-133796

COMPUTER TELECOMM SOC MODEL

525 LEFF, NATHANIEL H.
SOCIAL BENEFIT-COST ANALYSIS AND TELECOMMUNICATIONS INVESTMENT IN
DEVELOPING COUNTRIES
1984
INFORMATION ECONOMICS AND POLICY
1 N3 P217-226

TELECOMM ECON SOC FIN

526 LEFF, NATHANIEL H.
EXTERNALITIES, INFORMATION COST, AND SOCIAL BENEFIT-COST
ANALYSIS FOR ECONOMICS: AN EXAMPLE FROM TELECOMMUNICATION
1984
ECON AND CULTURAL CHANGE
32 N2 255-76

INFO COST ECON TELECOMM
SOC

527 LEGUYT, J.C. AND VISSER, D.A.
A DOMESTIC SATELLITE SYSTEM FOR PERU
1981, JUNE
PHILLIPS TELECOMMUNICATION REVIEW
39 N2 P71-82

PERU L AMER SAT TECH

528 LEGUYT, J.C., AND ROTS, I.B.J.
A DAMA SATELLITE SYSTEM FOR THIN-ROUTE TELEPHONE COUMMUNICATION
1983, JUNE
TELECOMMUNICATIONS REVIEW
41 N2 P134-144

SAT TELEPHONE RURAL TECHNICAL

529 LELONG, P., ABELIN, P. AND COTTEN, C.
LES TELECOMM EN AFRIQUE NOIRE ET A MADAGASCAR
1974 NOV
EUROPE OUTREMER
538

AF FRANCE RES MADAGASCAR

530 LENK, KLAUS
MICROELECTRONICS & SOCIETY. A REPORT TO THE CLUB OF ROME
FRIEDRICH, G. & SCHAFF, A. (EDS.)
1982
P261-296
MENTOR : NY

INFO COMPUTERS TECH ECON
POLIT SOC INTERNAT

531 LENT, J.A. (ED.)
BROADCASTING IN ASIA AND THE PACIFIC: A CONTINENTAL SURVEY OF RADIO AND TELEVISION
1978
TEMPLE UNIVERSITY PRESS: PHILADELPHIA, PA

ASIA PAC BROADCAST RADIO
TV HIST FIN

532 LERNER, DANIEL
THE PASSING OF TRADITIONAL SOCIETY
1958
FREE PRESS: GLENCOE, ILL.

SOC POLIT HIST

533 LERNER, DANIEL
TOWARD A COMMUNICATION THEORY OF MODERNIZATION: A SET OF CONSIDERATIONS. IN COMM AND POLITICAL DEVELOPMENT L. PYE (ED)
1963
CONFERENCE ON COMMUNICATIONS AND POLITICAL DEVELOPMENT
PRINCETON UNIVERSITY PRESS: PRINCETON, N.J.

THEORY

534 LERNER, DANIEL AND SCHRAMM, WILBUR (EDS)
COMMUNICATION AND CHANGE IN DEVELOPING COUNTRIES
1967
EAST-WEST CENTER: HONOLULU, HAWAII

MASS MEDIA INTERNAT POLICY TELECOMM

535 LESSER, B. & OSBERG, L.
THE SOCIO-ECONOMIC DEVELOPMENT BENEFITS OF TELECOMMUNICATION
1981
INTERNAT TELECOMM UNION: GENEVA

SOC ECON TELECOMM

536 LESSER, B. AND OSBERG,L.
THE IMPORTANCE OF TELECOMMUNICATIONS TO REGIONAL ECONOMIC DEVELOPMENT
1978
PHASE I REPORT: GOVERNMENT STUDIES PROGRAM
DALHOUSIE UNIVERSITY: HALIFAX

TELECOMM ECON

537 LESSER, BARRY
METHODOLOGICAL ISSUES INVOLVED IN ASSESSING THE ECONOMIC IMPACT OF TELECOMM WITH SPECIAL REFERENCE TO ISOLATED & UNDERPRIVILEGED AREAS
1978
DALHOUSIE UNIVERSITY: HALIFAX

TELECOMM ECON REMOTE THEORY

538 LESSER, BARRY
ON ASSESSING THE BENEFITS OF TELECOMMUNICATIONS: SOME PROPOSED METHODOLOGIES
1978, DEC 18-20
EXPERT MEETING ON TELECOMMUNICATION AND DEVELOPMENT
PARIS, FRANCE

TELECOMM ECON THEORY

539 LEVINE, ARTHUR L.
MANAGING SATELLITE TECHNOLOGY PROJECTS IN DEVELOPING COUNTRIES:
PROBLEMS AND SOLUTIONS
1981, NOVEMBER
IEEE ENGINEERING MANAGEMENT CONFERENCE
P179-182

SAT TECH MGT ECON
FIN

540 LEVY, MARION J., JR.
MODERNIZATION & THE STRUCTURE OF SOCIETIES
A SETTING FOR INTERNATIONAL AFFAIRS
1966
PRINCETON UNIVERSITY PRESS: PRINCETON, NEW JERSEY

SOC HIST

541 LEWIN, LEONARD (ED)
TELECOMM IN THE US: TRENDS & POLICIES
1981
ARTECH HOUSE: DEDHAM, MA.

TELECOMM US POLICY EVAL

542 LIBOIS, L.J.
VERS UNE APPROCHE GLOBALE DU PROBLEME DES TELECOMMUNICATIONS
1976
REVUE DES PTT

TELECOMM

543 LOCKSLEY, GARETH
THE POLITICAL-ECONOMY OF SATELLITE BUSINESS
1983, SEPT
TELECOMM POLICY
7, N3 195-203

INTERNAT SAT POLIT ECON
UK US

544 LONG, PETER
THE COMMUNICATION SATELLITE AS AN EDUCATIONAL TOOL
1982
CONVERGENCE
15 N1

SAT EDUC

545 LONNSTROM, S. AND MOO, I.
ECONOMIC ASPECTS IN TELEPHONE SYSTEMS
IN POLISHUK & O'BRYANT, EDS, TELECOMM & ECON DEVELOPMENT
1977
HORIZON HOUSE INTL: DEDHAM, MA

ECON TELEPHONE

546 LORENZI, NANCY M. ET AL
TELECOMMUNICATIONS INFORMATION NETWORK: A MODEL FOR ON-DEMAND
TRANSFER OF MEDICAL INFORMATION FINAL REPORT
ED239578
ERIC DOC REPRO SERV

US TELECOMM COMPUTER HEALTH
RURAL

547 LOVE, LEE
UNISPACE 82- A SUCCESSFUL FORUM? NEW COMMUNICATION TECHNOLOGY AND
THEIR IMPACT ON WESTERN INDUSTRIALIZED COUNTRIES
1984
27-32
FRIEDRICH EBERT STIFTUNG: BONN

INTERNAT POLICY UN SAT

548 MABOUCHE, ROBERT AND HADJTARHANI, NAIMEH
SEYCHELLES: ECONOMIC MEMORANDUM
1980
WORLD BANK: WASHINGTON, D.C.

AF SEYCHELLES ECON SOC
WORLD BANK RURAL

549 MACBRIDE COMMISSION
MANY VOICES,ONE WORLD: REPORT BY INTERNAT COMMISSION FOR THE STUDY OF COMMUNICATION PROBLEMS
1980
UNESCO: PARIS

WORLD INTERNAT TELECOMM EVAL

550 MADDICK, HENRY
DEMOCRACY, DECENTRALIZATION AND DEVELOPMENT
1963
ASIA PUBLISHING HOUSE: NEW YORK

POLIT ECON

551 MADDOX, BRENDA
THE BORN-AGAIN TECHNOLOGY
1981, AUG 27
THE ECONOMIST
P3-21

TECH ECON

552 MAGNANT, R.S.
SATELLITE COMMUNICATIONS: US POLICY CONSIDERATIONS
DEC 1979
POLICY (GB)
3 N4 297-304

SAT POLICY US

553 MAHON, F.V.
PERSPECTIVES OF ACCESS TO ON-LINE INFO BY THE DEVELOPING COUNTRIES
1982 OCT-DEC
UNESCO J OF INFO SCI, LIBRARIANSHIP & ARCHIVES ADM
4 N4
ERIC #EJ276799

COMPUTER INFO UNESCO

554 MAITLAND COMMISSION
THE MISSING LINK: REPORT OF THE INDEPENDENT COMMISSION FOR
WORLDWIDE TELECOMMUNICATIONS DEVELOPMENT
1984
INTERNATIONAL TELECOMMUNICATION UNION
ITU: GENEVA

WORLD TELECOMM

555 MAITLAND, SIR DONALD
INDEPENDENT COMMISSION FOR WORLDWIDE TELECOMMUNICATIONS DEVELOPMENT
1985
TELEMATICS AND INFORMATICS
2 N1 P97-100

TELECOMM WORLD

556 MALECELA, J.C.
COMMUNICATIONS POLICY AND INFRASTRUCTURE DEVELOPMENT
1983
PAPER/ TANZANIA POSTS AND TELECOMM CORP.
DAS ES SALAAM, TANZANIA/ITU TELECOMM FORUM: GENEVA

INFO TELECOMM URBAN RURAL
POLICY

557 MALECKI, IGNACY
THE ROLE OF SCIENTIFIC-TECHNICAL REVOLUTION IN DEVELOPMENT
1980
SCIENCE OF SCIENCE
1 N1 P67-77

TECH

558 MALONEY, E.
GREEN THUMB FARM INFORMATION PROJECT TAKES ROOT IN STATE
OF KENTUCKY
1982, JAN 25
MICROCOMPUTING
P341

US RURAL TELECOMM COMPUTER
AGRIC VTTEXT

559 MANSELL, ROBIN E.
ADAPTATION VS TRANSFORMATION:THE IMPACT OF THE NEW PARADIGM IN THE COMMUNICATION AND DEVELOPMENT FIELD
1982, JUNE
CANADIAN J OF COMMUNICATION
8 N3 P42-60

POLICY POLIT ECON TECH
THEORY

560 MARRIS, M.D. (ED)
TOWARD A NEW WORLD INFORMATION ORDER?
1981/82 FALL/WI
JOURNAL OF INTERNAT AFFAIRS
35, N2

INTERNAT INFO POLIT CULT
SOC UNESCO POLAND PAKISTAN

561 MARSH, DONALD
TELECOMMUNICATIONS AS A FACTOR IN THE ECONOMIC DEVELOPMENT OF A COUNTRY
1976
IEEE TRANSACTIONS ON COMM
24 P716-22

TELECOMM ECON POLIT

562 MARTIN, RICHARD
THE PERU RURAL COMMUNICATION SERVICES PROJECT
1981,JUNE
UPLINK
P2-3

PERU L AMER RURAL

563 MARZOUKI, ABELKADAR
URTNA: BROADCAST IN AFRICA
1981
INTERMEDIA
9:5

PANAFTEL AF BROADCAST SAT

564 MASMOUDI, M.
THE NEW WORLD INFORMATION ORDER
1979, SPRING
JOURNAL OF COMMUNICATION
24 N2 P172-185

INFO POLICY POLIT INTERNAT

565 MASON, ROBERT A.
COMMUNICATION ANALYSIS FOR PILOT PROJECT TO IMPROVE PREVENTIVE RURAL HEALTH DELIVERY: EGYPT
1976, JUNE
AMERICAN EDUCATIONAL DEVELOPMENT: WASHINGTON, D.C.

EGYPT RURAL HEALTH

566 MASONER, L.M.,MASONER, P.M., BERNAL, H.
AN EXPERIMENT IN RADIOPHONIC EDUCATION CULTURE POPULAR
1982
PROSPECTS
12 N3 P365-374

RADIO EDUC CULT

567 MATTA, FERNANADO REYES
WORLD COMMUNICATIONS. IN GERBNER, GEORGE, & SIEFERT, MARSHA (EDS.)
1984

INTERNAT INFO TECH SOC
ECON POLIT RES THEORY

568 MATTA, HAROLD CORREA
TECHNOLOGY AND DEVELOPING COUNTRIES
1980
TELECOMMUNICATION JOURNAL
47 N8 P459-461

BRAZIL TECH TELECOMM RES

569 MATTELART, ARMAND AND SIEGLAUB, SETH, EDS.
COMMUNICATION AND CLASS STRUGGLE: CAPITALISM AND IMPERIALISM
1979
1 P23-71
INTERNATIONAL GENERAL: NY

MASS MEDIA SOC ECON POLIT
THEORY

570 MATTELART, ARMAND AND SIEGLAUB, SETH
COMMUNICATION AND CLASS STRUGGLE
1983
2 P413-418
INTERNATIONAL GENERAL: NY

TECH POLIT ECON SOC
INTERNAT THEORY

571 MATTELART, ARMAND AND SCHMUCLER, HECTOR
COMMUNICATION AND INFORMATION TECHNOLOGY: FREEDOM OF CHOICE FOR LATIN AMERICA
1985
ABLEX: NORWOOD, N.J.

COMPUTER INFO POLIT ECON
L AMER

572 MATTELART, MICHELE AND MATTELART, ARMAND
SMALL TECHNOLOGY: THE CASE OF MOZAMBIQUE
1980, SPRING
JOURNAL OF COMMUNICATION
32 N2

MOZAMBIQUE TECH

573 MAXMEN, J.S.
TELECOMMUNICATIONS IN PSYCHIATRY
1978, JULY
AMERICAN JOURNAL OF PSYCHOTHERAPY
32, N3 P450-456

TELECOMM HEALTH TRAIN TV
TELEPHONE

574 MAYO, J.K. AND HORNIK, R.C.
EVALUATION AND RESEARCH IN THE PLANNING, DEVELOPMENT AND SUPPORT OF MEDIA BASED ON EDUCUCATION
1980
INTERNAT INSTITUTE FOR EDUCATION PLANNING : PARIS

EDUC PLAN EVAL

575 MAYO, J.K. AND SPAIN, P.L.
COMMUNICATION POLICY AND PLANNING FOR EDUCATION AND DEVELOPMENT
1977, MAY
THE INSTITUTE FOR COMMUNICATION RESEARCH: STANFORD UNIVERSITY

POLICY PLAN EDUC INDIA
TANZANIA EL SALVADOR KOREA KENYA
ALASKA

576 MAYO, JOHN S.
THE MICROELECTRONICS REVOLUTION: THE BENEFITS OF HIGH TECHNOLOGY SPREAD BEYOND THE RICHER NATION
1985
THIRD WORLD DEVELOPMENT
2 P181-182

TELECOMM

577 MCANANY, E.G. AND MAYO, J.K.
COMMUNICATIONS MEDIA IN EDUCATION FOR LOW INCOME COUNTRIES: IMPLICATIONS FOR PLANNNING
1980
UNESCO PRESS: PARIS

EDUC PLAN MEDIA

578 MCANANY, E.G., OLIVEIRA, J.O., ORIVEL, F., STONE, J.
DISTANCE EDUCATION: NEW APPROACHES IN EDUCATION FOR DEVELOPING COUNTRIES
1983
EVALUATION IN EDUCATION
6 P389-376

MEDIA EDUC REMOTE

579 MCANANY, EMILE G.
RADIO'S ROLE IN DEVELOPMENT: FIVE STRATEGIES OF USE.
1973
ACADEMY FOR EDUCATIONAL DEVELOPMENT: WASHINGTON, D.C.

RADIO

580 MCANANY, EMILE G.
THE EXERN PROJECT IN BRAZIL: AN ANALYTICAL CASE STUDY
1980
UNESCO PRESS: PARIS, FRANCE

BRAZIL L AMER

581 MCANANY, EMILE G. (ED.)
COMMUNICATIONS IN THE THIRD WORLD: THE ROLE OF INFORMATION IN DEVELOPMENT
1980
PRAEGER PUBLISHERS: NEW YORK

AGRIC HEALTH SOC POLIT
RURAL HIST

582 MCANANY, EMILE G. AND OLIVEIRA, J.
THE SACI/EXTERM PROJECT IN BRAZIL: AN ANALYTICAL CASE STUDY
1980
REPORTS AND PAS. ON MASS COMM
N89
UNESCO: PARIS

UNESCO BRAZIL L AMER SAT
EDUC RURAL

583 MCANANY, EMILE G. ET AL.
DISTANCE EDUCATION FOR DEVELOPING COUNTRIES
1983
PERGAMM PRESS LTD: GREAT BRITAIN

EDUC REMOTE

584 MCCRONE, D.J. & CNUDDE,C. F.
TOWARD A COMMUNICATIONS THEORY OF DEMOCRATIC POLITICAL DEVELOPMENT:A CAUSAL MODEL
1967
AMER POLIT SCIENCE REVIEW
61 P72-9

THEORY POLIT MODEL

585 MCGILLEM, CLARE D., MCLAUCHLAN, WILLIAM P.
HERMES BOUND: THE POLICY & TECHNOLOGY OF TELECOMMUNICATION
1978
SERIES:SCIENCE & SOCIETY
2
PURDUE UNIV

TELECOMM POLICY TECH

586 MCGREGOR, ALAN
THIRD WORLD LINKS, PART I: TRAUMATIC ENCOUNTERS WITH THE PHONE SYSTEM
1985, APRIL 29
THE TIMES OF LONDON

TELEPHONE

587 MCGREGOR, ALAN
THIRD WORLD LINKS, PART II: SATELLITE LIFELINE TO GLOBAL VILLAGE
1985, APRIL 30
THE TIMES OF LONDON

SAT

588 MCKNIGHT, L
THE DEREGULATION OF INTERNATONAL SATELLITE COMMUNICATION: US POLICY & THE INTELSAT RESPONSE
1985 MARCH
SPACE COMMUN & BROADCAST (NETHERLANDS)
3 N1 P39-59

SAT DEREG INTELSAT POLICY
INTERNAT US

589 MCLELLAN, IAIN
TELEVISION FOR DEVELOPMENT: THE AFRICAN EXPERIENCE
1986
INTERNATIONAL DEVELOPMENT RESEARCH CENTRE: OTTAWA, CANADA

AF TV

590 MCPHAIL, THOMAS
ELECTRONIC COLONIALISM: THE FUTURE OF INTERNATIONAL BROADCASTING AND COMMUNICATION
1981
SAGE PUBLISHING COMPANY: BEVERLY HILLS, CA.

INTERNAT TELECOMM MASS MEDIA POLICY
ECON ITU UNESCO

591 MEDNIKOV, D.
WHAT IS THE EFFECT OF DISPATCHER COMM IN FARMING
KAKOJ EFFEKT DACTVEDRENIE DISPETCERSKOJ SVJAZI...
1975
VESTNIK SVJAZI
4

TELECOMM AGRIC RURAL

592 MEED, J.
THE USE OF RADIO IN OPEN UNIVERSITY COURSE DESIGN 1971-1974
1976
SURVEY/ MILTON KEYNES, OPEN UNIVERSITY

RADIO EDUC

593 MEIER, R.
A COMMUNICATION THEORY OF URBAN GROWTH
1962
MIT PRESS: CAMBRIDGE, MA.

THEORY URBAN ECON

594 MELODY, WILLIAM H, (ED)
THE INFORMATION SOCIETY: UNVEILING SOME CONTRADICTIONS
1985, JULY
MEDIA CULTURE AND SOCIETY
7 N3

INFO	ECON	POLICY	INTERNAT
POLIT	EUR	L AMER	CANADA

595 MELODY, WILLIAM H, SALTER, LIORA, AND HEYER, PAUL (EDS)
CULTURE, COMMUNICATION & DEPENDENCY
1981
TELECOMMS

RES	THEORY	SOC	ECON
CANADA			

596 MELODY, WILLIAM H.
ARE SATELLITES THE PYRAMIDS OF THE 20TH CENTURY?
1979, SPRING
IN SEARCH
P4-8

SAT	POLICY	SOC	ECON
REG	CANADA		

597 MELODY, WILLIAM H.
PERSPECTIVES IN COMMUNICATION POLICY & PLANNING
IN SYED RAHIM (ED)
1977
P25-42
EAST-WEST COMMUNICATION INSTITUTE: HONOLULU, HAWAII

INTERNAT	PLAN	POLICY	ECON
PLAN			

598 MELODY, WILLIAM H.
TELECOMMUNICATIONS IN ALASKA: ECONOMICS AND PUBLIC POLICY
1978
PB-297 8880
GOVERNOR'S OFFICE OF COMMUNICATION: ANCHORAGE, ALASKA

TELECOMM	ALASKA	ECON	POLICY

599 MELODY, WILLIAM H.
THE INFORMATION SOCIETY: IMPLICATIONS FOR ECONOMIC
INSTITUTIONS AND MARKET THEORY
1985
JOURNAL OF ECONOMIC ISSUES

INFO POLICY ECON INTERNAT

600 MELODY, WILLIAM H.
THE ROLE OF COMMUNICATION IN DEVELOPMENT PLANNING. IN PERSPECTIVES
ON COMM POLICY AND PLANNING. RAHIM, S.A. AND MIDDLETON, JOHN (EDS)
1977
MONOGRAPH
N3
EAST-WEST COMM INST: HONOLULU, HAWAII

PLAN

601 MELROSE, ENRIQUE
THE NEED FOR TELEMATICS POLICIES IN DEVELOPING COUNTRIES
1983
PACIFIC TELECOMMUNICATIONS CONFERENCE
HONOLULU, HAWAII

T-MATICS POLICY SOC ECON
TECH

602 MERRITT, ROBERT P.
THE ALASKA TELECONFERENCING SYSTEM
UNDATED
UNIVERSITY OF ALASKA: FAIRBANKS, ALASKA

ALASKA TELECONF TELECOMM REMOTE
TV TELEPHONE BUSI EDUC

603 MESEL, E. AND WIRTSCHAFTER, D.D.
AUTOMATION OF A PATIENT MEDICAL PROFILE FROM INSURANCE CLAIMS DATA:
A POSSIBLE FIRST STEP IN AUTOMATING AMBULATORY MEDICAL RECORDS
1976, WINTER
HEALTH AND SOCIETY
P29-45

HEALTH

604 MEYER, RICHARD J.
AFGHANISTAN. TRANSITION TO TELEVISION
1978
PUBLIC TELECOMMUNICATIONS REVIEW
6 N1 P45-47

AFGHANISTAN TV RADIO

605 MGAYA, F.M.
ECONOMIC CONSIDERATIONS IN PLANNING TELECOMMUNICATION DEVELOPMENT IN ISOLATED AND UNDERDEVELOPED AREAS
1978

ECON TELECOMM PLAN RURAL

606 MICHELON, RICHARD L.
APPROPRIATE SYSTEMS FOR RURAL TELECOMMUNICATION
1982
PAPER/ GLOBECOM 82: MIAMI, FL

TELECOMM TECH RURAL

607 MICHELSON, RICHARD L.
COMMUNICATION SATELLITES FOR RURAL SERVICE: PROBLEMS AND POTENTIALS
1981, MARCH
TELECOMM POLICY

SAT RURAL

608 MIDDLETON, JOHN & LIM, M.K.
ADVENTURES IN SIABAD: A SIMULATION-GAME WORKSHOP FOR COMMUNICATION PLANNING
1979, JAN
PAPER PRESENTED AT PAC TELECOMM CONFER

PLAN

609 MIDDLETON, JOHN AND LIM, M.K.
ADVENTURES IN SIABAD: A SIMULATION GAME WORKSHOP FOR COMMUNICATION PLANNING
1979
PROCEEDINGS OF PACIFIC TELECOMM CONFERENCE: HONOLULU, HI

PLAN RES TELECOMM

610 MILI, MOHAMMED
PANAFTEL AND PAN AFRICAN TELECOMMUNICATION UNION, EDITORIAL
1976 APR
TELECOMM JOURNAL
43 N7

AF PANAFTEL ITU SAT

611 MILI, MOHAMMED
THE ROLE OF THE ITU IN AFRICAN TELECOMMUNICATION
1978, DEC
AFRICA (AFRICA JOURNAL, LTD)
P48

AF ITU POLICY TELECOMM

612 MILI, MOHAMMED
WHY TELECOMM IS ESSENTIAL FOR SOCIAL/ECONOMIC DEVELOPMENT
1980, JUNE 15
TELEPHONE ENGINEER & MGMT
84, N12 P57-61

TELECOMM SOC ECON

613 MILI, MOHAMMED (ED.)
SPEAKER'S PAPERS: THIRD WORLD TELECOMMUNICATION FORUM: EXECUTIVE SYMPOSIUM ON TELECOMM PERSPECTIVES AND ECONOMIC IMPLICATIONS
1980
INTERNATIONAL TELECOMMUNICATION UNION
GENEVA, SWITZERLAND

INTERNAT TELECOMM COMPUTER SOC
ECON TECH FIN

614 MILLER, A.D.D.
THE ROLE OF SATELLITE COMMUNICATIONS IN A DOMESTIC TELECOMMUNICATION ENVIRONMENT
MILLER COMMUNICATION SYSTEMS, LIMITED
KANATA, ONTARIO

SAT TELECOMM TECH

615 MILLS, MIRIAM K.
COMPUTER CONFERENCING FOR THE THIRD WORLD: ROADMAP FOR THE FUTURE

TELECONF COMPUTER TECH ECON

616 MISHAN, E.J.
THE POSTWAR LITERATURE ON EXTERNALITIES: AN INTERPRETIVE ESSAY
1971
JOURNAL OF ECONOMIC LITERATURE
9 P1-28

ECON SOC

617 MISHAN, E.J.
THE RELATIONSHIP BETWEEN JOINT PRODUCTS, COLLECTIVE GOODS AND EXTERNAL EFFECTS
1969
JOURNAL OF POLITICAL ECON
77 P329-48

ECON

618 MITCHELL, W.C.
THE USE OF SATELLITES IN MEETING THE TELECOMMUNICATIONS NEEDS OF DEVELOPING COUNTRIES
1975
TECH REP #1

SAT TELECOMM

619 MIYAUCHI,K.
DOMESTIC SATELLITE COMMUNICATION SYSTEM RESEARCH AND DEVELOPMENT
JULY/AUG 1980
REV. ELECTR. COMMUN. LAB. (JAPAN)
28 V7-8

JAPAN ASIA RES TECH
SAT

620 MODY, BELLA
LESSONS FROM THE INDIAN SATELLITE EXPERIMENT
1978, SEPTEMBER
EDUCATIONAL BROADCASTING INTERNATIONAL
11 N3

INDIA SAT PLAN

621 MOEMEKA, ANDREW A.
AFRICA AND THE RIGHT TO COMMUNICATE
1979, JANUARY
AFRICA
P54-55

AF RADIO TV RURAL
URBAN

622 MOENANDIR, WILLY
INDONESIAN TELECOMMUNICATIONS DEVELOPMENT: AN OVERVIEW
1985
TELEMATICS AND INFORMATICS
2 N1 P79-89

INDONESIA TELECOMM ECON POLIT

623 MOENANDIR, WILLY
INDONESIAN TELECOMMUNICATIONS DEVELOPMENT: ITS ROLE IN THE
PURSUIT OF NATIONAL DEVELOPMENT'S OBJECTIVE
1979
PROCEEDINGS OF PACIFIC TELECOMM CONFERENCE: HONOLULU, HI

INDONESIA TELECOMM ECON POLIT
AGRIC BUSI JAKARTA

624 MOLINO, ENZO
DATA BASES: CONSIDERATIONS OF RELEVANCE TO DEVELOPING COUNTRIES
1982, OCT-DEC
UNESCO J OF INFO SCI, LIBRARIANSHIP & ARCHIVES ADM
4 N4 245-252

DATA INFO COMPUTER INTERNAT

625 MONSON, MAVIS
BRIDGING THE DISTANCE
1978
REGENTS OF THE UNIVERSITY OF WISCONSIN SYSTEM: MADISON, WI

RES EDUC TELECONF EVAL

626 MONTELLA, L.
GLOBAL COMPETITION AND EEC POLICY IN TELECOMMUNICATION; AN ITALIAN PERSPECTIVE
1984, SEPT
TELECOMM POLICY
8 N3 P 205

ITALY EUR POLICY INTERNAT
POLIT

627 MONTEMAYOR, JESUS M.
GRASSROOTS INFORMATION SYSTEM AND INTEGRATED RURAL DEVELOPMENT
1983, FEB 23-24
PAPER/ DEVELOPMENT FORUM, UNITED NATIONS, SWITZERLAND

INFO RURAL SOC

628 MONTGOMERY, HALE
ALGERIA EXEMPLIFIES TELECOMMUNICATIONS IN DEVELOPING COUNTRIES
1981, JULY
SATELLITE COMMUNICATIONS
P16

ALGERIA TELECOMM

629 MONTMANEIX, M.G.
LE TELEPHONE
1974
QUE SAIS-JE
251
PUF:PARIS

TELEPHONE

630 MOORE, DONALD GARY
THE IMPORTANCE OF TELECOMMUNICATION TO A MODERN LATIN AMERICA
1982
MASTERS ABSTRACTS
20/04 P373

L AMER TELECOMM EVAL

631 MORGAN, W.L.
THE NEXT DECADE. AN ECONOMIC OUTLOOK FOR THE EIGHTIES. (SATELLITE COMMUNICATIONS)
1981, JAN
SATELLITE COMMUNICATIONS
5 N1 P20-6

SAT ECON

632 MORGANTI, G., PALADIN, G., SAVINO, A.
THE SOCIO-ECONOMICAL IMPACT DUE TO SOME NEW BASIC TELECOMMUNICATION SERVICES
1983, AP-JUN
REV. FITCH (BELGIUM)
22 N2 P19-24

TELECOMM SOC ECON

633 MOSCO, VINCENT & HERMAN, ANDREW
COMMUNICATION & SOCIAL STRUCTURE: CRITICAL STUDIES IN MASS MEDIA RESEARCH. IN MCANANY, SCHNITMAN, & JANUS (EDS.)
1981
P58-84
PRAEGER PUB: NY

THEORY SOC POLIT INTERNAT

634 MOSCO, VINCENT; VOIGT, MELVIN J.(ED)
POLICY RESEARCH IN TELECOMMUNICATION: PROCEEDINGS, TELECOMM POLICY RESEARCH CONFER ANNUAL 11TH
1984
ABLEX: NORWOOD, N.J.

TELECOMM RES POLICY

635 MOSS, MITCHELL (ED.)
TELECOMMUNICATIONS AND PRODUCTIVITY
1981
ADDISON WESLEY: READING, MA.

THEORY TELECOMM RES ECON
SOC US EVAL INFO

636 MOSS, P.R.
THE DEVELOPMENT OF GLOBAL SATELLITE TELECOMMUNICATION
1983, FEB
J. BR. INTERPLANET SOC (GB)
36 N2 P66-72

SAT TELECOMM TECH WORLD

637 MOUL, B.A.
ISDN-DEVELOPMENTS IN ASIA
1985, JAN 13-16
PAPER/ PACIFIC TELECOMMUNICATION CONFERENCE
P244-254

PAC ASIA TELECOMM DIGITAL
INFO

638 MOUSA, MOUSA MOHAMED
ROLE OF TRUNKED RADIO TELEPHONE TECHNOLOGY IN RURAL COMMUNICATION NETWORKS
1980
DISSERTATION ABSTRACTS INTERNAT
41/02-B P642

TELEPHONE 2-WAY RADIO TECH RURAL

639 MUIS, A.
SOME IMPLICATIONS OF TELEVISION EXPOSURE AMONG TRADITIONAL PEASANTS: A CASE FROM SOUTH SULAWESI ISLAND
1983
MEDIA ASIA
10 N2 P92-100

INDONESIA ASIA TV RURAL

640 MUQBIL, I.
ASIAN PLANNERS LOOK TO SATELLITES IN SEARCH OF SOLUTIONS
1984, APR
COMPUT ASIA (HONG KONG)
N66 P40-2

ASIA PLAN SAT

641 MURPHY, BRIAN
THE WORLD WIRED UP: UNSCRAMBLING THE NEW COMMUNICATIONS PUZZLE
1983
COMEDIA PUBLISHING GROUP: LONDON

WORLD SAT CABLE TV
INFO EVAL

642 MYAGA, FRANK M.
ECONOMIC CONSIDERATIONS IN PLANNING TELECOMMUNICATIONS DEVELOPMENT IN ISOLATED AND UNDERPRIVILEGED AREAS
1978
TANZANIA POSTS AND TELECOMMUNICATIONS CORPORATION

TELECOMM ECON PLAN REMOTE

643 MYRDAL, G.
THE CHALLENGE OF WORLD POVERTY
1970
PENGUIN: NEW YORK

WORLD ECON POLICY POLIT

644 NARAYANAN, K., AND VENUGOPAL, D.
BENEFITS AND FEATURES OF CERTAIN APPLICATIONS OF SATELLITE COMMUNICATIONS FOR DEVELOPING COUNTRIES
1985, MAY
TELECOMMUNICATION JOURNAL
52 N6 P298-303

SAT RURAL BUSI COMPUTER
DATA

645 NATIONAL COUNCIL OF APPLIED ECONOMIC RESEARCH
SURVEY OF RURAL PUBLIC CALL OFFICES
1978

RURAL TELEPHONE STAT

646 NEGROTTI, MASSIMO
CULTURAL DYNAMICS IN THE DIFFUSION OF INFORMATICS
1984, FEB
FUTURES
16, N1 P38-46

INFO COMPUTERS CULTURE TECH

647 NEILSON, BOB
A WORKSHOP ON DISTANCE EDUCATION IN CHINA
1981, SEPTEMBER
EDUCATIONAL BROADCASTING INTERNATIONAL
14 N3 P 135-139

CHINA EDUC

648 NICKELSON, R.L. & TUSTISON, G.F.
AN EARTH STATION DESIGN FOR RURAL TELECOMM
MIMEO
INTERNAT TELECOMM UNION: GENEVA

SAT RURAL TELECOMM

649 NICOL, L.Y. AND TEITZ, MICHAEL
ECONOMIC AND SPATIAL EFFECTS OF INVESTMENT IN TELECOMMUNICATIONS IN DEVELOPING COUNTRIES: A CONCEPTUAL ANALYSIS (ARTICLE)
1981
FISHER, C.(ED.): THE EFFECTS OF TELECOMM: A RESEARCH PERSPECTIVE
UNIVERSITY OF CALIFORNIA: BERKELEY

TELECOMM ECON FIN

650 NICOL, LIONEL Y.
COMMUNICATIONS, ECONOMIC DEVELOPMENT, AND SPATIAL STRUCTURES; A REVIEW OF RESEARCH
1983, MAY
INSTITUTE OF URBAN AND REGIONAL DEVELOPMENT, UNIVERSITY OF CA.
UNIVERSITY OF CALIFORNIA: BERKELEY, CA.

TELECOMM FIN ECON SOC
RES THEORY EVAL

651 NIE, N.,HULL, C., JENKINS, J., STEINBRENNER, K., & BENT, D.
STATISTICS PACKAGE FOR THE SOCIAL SCIENCES
1975
2ND ED
MC GRAW-HILL: NEW YORK

SOC STAT

652 NNYIRA, JUMA E.
THE ROLE OF EDUCATIONAL RADIO IN NATIONAL DEVELOPMENT: A REVIEW OF EXPERIENCES IN DEVELOPING COUNTRIES
1980-81
INTERNAT AND INSTRUCTIONAL MEDIA
8 N3 P255-260

RADIO EDUC RURAL

653 NORA, S., AND MINC, A.
THE COMPUTERIZATION OF SOCIETY
1980
MIT PRESS: CAMBRIDGE, MA

COMPUTER SOC

654 NORDENSTRENG, KAARLE
U.S. POLICY AND THE THIRD WORLD: A CRITIQUE
1982, SUMMER
J OF COMMUNICATION
32, N3 P54-59

INTERNAT TELECOMM MASS MEDIA US
POLICY

655 NORDENSTRENG, KAARLE AND SCHILLER, HERBERT (EDS)
NATIONAL SOVEREIGNTY AND INTERNATIONAL COMMUNICATION
1979
ABLEX PUBLISHING: NORWOOD, N.J.

INTERNAT POLIT ECON TECH
POLICY

656 NORIEGA, L. A. AND LEACH, F.
BROADCASTING IN MEXICO
1979
ROUTLEDGE AND KEGAN PAUL LTD: LONDON

MEXICO BROADCAST CULT ECON
POLIT

657 NORRIS, WILLIAM C.
THE PRIVATE SECTOR'S ROLE IN THE THIRD WORLD
1985
THIRD WORLD DEVELOPMENT
2 P129-133

TECH SOC FIN

658 NORWOOD, FRANK W.
PUBLIC TELECOMMUNICATION POLICIES AND EDUCATION'S OPTIONS
UNDATED
JOINT COUNCIL ON EDUCATIONAL TELECOMMUNICATIONS: WASHINGTON, D.C.

SAT POLICY EDUC TV
POLICY TECH

659 NOYELLE, THIERRY J.
WORK IN A WORLD OF HIGH TECH: PROBLEMS AND PROSPECTS FOR ECONOMICALLY DISADVANTAGED WORKERS. STATE-OF-THE-ART PAPER
ED254617
ERIC DOC REPRO
TECH

EDUC TRAIN SOCIAL

660 NYIRENDA, JUMA ESAU
THE ROLE OF EDUCATIONAL RADIO IN NATIONAL DEVELOPMENT: A REVIEW OF EXPERIENCES IN DEVELOPING COUNTRIES
1980-81
INTERNATIONAL INSTITUTIONAL MEDIA
8 N3 P255-260

RADIO EDUC ECON SOC
CULT POLIT

661 O'NEIL, PAUL (ED.)
INTERNATIONAL TELECOMMUNICATION TRADE REPORTS
1978, DEC 30
INFORMATION GATEKEEPERS, INC: BROOKLINE, MA

COSTA RICA PARAGUAY URUGUAY ECUADOR
BOLIVIA FIN

662 O'SULLIVAN-RYAN, J.O.
TELECOMMUNICATIONS PLANNING IN VENEZUELA. IN PROCEEDINGS/TENTH ANNUAL TELECOMMUNICATIONS POLICY RESEARCH CONFERENCE
1983
P321-334
ABLEX PUBLISHERS: NORWOOD, NEW JERSEY

VENEZUELA TELCOMM PLAN TELEPHONE
TECH RES

663 OFFICE OF SCIENCE & TECH, WASH D.C.
SUMMARY OF SURVEY ON THE FEDERAL ROLE IN SATELLITE COMMUNICATION RESEARCH AND DEVELOPMENT
1978, JULY
PB-295 949/2

US POLICY REG SAT
RES

664 OFFICE OF TELECOMM POLICY
INTERAGENCY REPORT ON RURAL COMMUNICATIONS
1977
OTA: WASHINGTON,D.C.

RURAL

665 OGAN, C.L.
MEDIA DIVERSITY & COMMUNICATION POLICY: IMPACT OF VCRS & SATELLITE TV
1985 MARCH
TELECOMM POLICY
9 N1 63-73

SAT TV EVAL RES
MEDIA POLICY

666 OHLMAN, HERBERT
TECHNICAL FEASIBILITY OF USING SATELLITE BASED COMMUNICATIONS IN SUPPORT OF HEALTH PROGRAMS IN WESTERN PACIFIC REGION
1980, MARCH
WORLD HEALTH ORGANIZATION: GENEVA, SWITZERLAND

WESTERN PAC SAT HEALTH

667 OHLMAN, HERBERT
TELECOMMUNICATION TRANSPORT TRADEOFFS
1981
INTERNATIONAL TELECOMMUNICATION UNION: GENEVA

TELECOMM

668 OKOLIE, C.C.
LEGAL ASPECTS OF THE USE OF THE GEOSTATIONARY ORBIT & THE INTEREST OF AFRICAN STATES IN SATELLITE TELECOMM
1984, JUNE
SPACE COMM & BROADCASTING (ISSN 0167-9368)
2

AF SAT POLICY LEGAL

669 OKUNDI, P., OGAWE, A.W., AND KIBOMBO, J.P.
RURAL TELECOMMUNICATIONS DEVELOPMENT IN EAST AFRICA
IN POLISHUK & O'BRYANT, EDS, TELECOMM & ECON DEVELOPMENT
1977
HORIZON HOUSE INTL: DEDHAM, MA

TELECOMM RURAL EAST AF

670 OKUNDI, PHILLIP O.
PAN-AFRICAN TELECOMMUNICATION NETWORK: A CASE FOR TELECOMMUNICATIONS
IN DEVELOPMENT OF AFRICA
1976, JULY
IEEE TRANSACTIONS ON COMMUNICATIONS
COM24N7 P749-54

AF PANAFTEL TELECOMM ORBIT

671 OKUNDI, PHILLIP O.
THE INTEGRATION OF TELECOMMUNICATION SERVICES WITH PLANNING IN
OTHER SECTORS
1978
KENYA POSTS AND TELECOMMUNICATIONS CORPORATION

TELECOMM PLAN

672 OLMESDAHL, DERICK BASSON
THE HISTORY OF TELECOMMUNICATION IN SOUTH AFRICA 1873-1887
1980
MASTERS ABSTRACTS
20/02 P183

S AF AF HIST TELECOMM

673 ORGANIZATION FOR ECONOMIC COOPERATION AND DEVELOPMENT (OECD)
TELECOMMUNICATION: PRESSURES AND POLICIES FOR CHANGE
1983

TELECOMM POLICY

674 OSBORNE, W.P., SMITH, R.L., AND STAPON, H.J.
SUDOSAT: THE NATIONAL DOMESTIC SATELLITE COMMUNICATIONS SYSTEMS FOR THE GOVERNMENT OF THE DEMOCRATIC REPUBLIC OF SUDAN
1977
PAPER/ INTELCOM 77/ ATLANTA, GEORGIA
HORIZON HOUSE INTERNATIONAL: DEDHAM, MA

SUDAN SAT

675 OSEGUERA, A. ANTHONY
MEXICO: THE ROLE OF THE PARTICIPATORY MEDIA IN IMMIGRATION/ EMIGRATION AS CULTURAL AND POLITICAL ECONOMY
ED251268
ERIC DOC REPRO SERV

MEXICO MEDIA POLIT CULT
POLICY

676 OSLUND, ROBERT J.
PACIFIC FACILITIES PLANNING PROCESS: AN APPROPRIATE MECHANISM
COMMUNICATIONS SATELLITE CORPORATION: WASHINGTON, D.C.

PAC INTERNAT SAT CABLE
PLAN LEGAL

677 PAISLEY, W.
COMPUTERIZING INFORMATION: LESSONS OF A VIDEOTEXT TRIAL
1983
JOURNAL OF COMMUNICATION
33 N1 P153-161

V-TEXT US INFO COMPUTER
RURAL AGRIC

678 PARADA, NELSON DE JESUS
A BRIEF DESCRIPTION OF THE CURRENT STATUS AND FUTURE PLANS OF THE BRAZILIAN SATELLITE PROGRAM
1983, FEBRUARY
PAPER/FIRST INTERGOVERNMENTAL MEETING OF SPACE TECHNOLOGY EXPERTS
COLUMBIA UNIVERSITY: NEW YORK

BRAZIL L AMER SAT TECH

679 PARK, BEN
AN INTRODUCTION TO TELEMEDICINE
1974
THE ALTERNATIVE MEDIA CENTER, NEW YORK UNIVERSITY: NEW YORK

HEALTH TELECOMM

680 PARKER, E.B.
AN INFORMATION-BASED HYPOTHESIS
1978
JOURNAL OF COMMUNICATION
28 P81-3

THEORY INFO

681 PARKER, E.B.
APPROPRIATE TELECOMMUNICATIONS FOR ECONOMIC DEVELOPMENT
1984, SEPTEMBER
TELECOMMUNICATION POLICY
8 N3 P173-180

TELECOMM ECON SAT DIGITAL
TECH

682 PARKER, E.B.
COMMUNICATION SATELLITES FOR RURAL DEVELOPMENT
1978 MAY
PAPER PRESENTED TO INTERNAT SATELLITE COMM SEMINAR, LIMA PERU

RURAL SAT

683 PARKER, E.B.
ECONOMIC AND SOCIAL BENEFITS OF THE REA TELEPHONE LOAN PROGRAM
1981
INTERNATIONAL TELECOMMUNICATION UNION: GENEVA

ECON SOC EVAL FIN
RURAL TELEPHONE

684 PARKER, E.B.
PLANNING COMMUNICATION TECHNOLOGY & INSTITUTIONS FOR DEVELOPMENT
1976
PAPER FOR EAST-WEST CENTER ON COMM POLICY & PLANNING FOR DEVELOPMENT
EAST-WEST CENTER: HONOLULU, HAWAII

TELECOMM TECH PLAN

685 PARKER, E.B. & LUSIGNAN, B.
TECHNICAL AND ECONOMIC CONSIDERATIONS IN PLANNING
RADIO SERVICES
1977
IN RADIO FOR EDUCATION & DEVEL.: CASE STUDIES, SPAIN, ET AL (EDS)
2 P443-459
WORLD BANK: WASHINGTON, D.C.

RADIO PLAN TECHNICAL ECON

686 PARKER, E.B. AND MOHAMMDI, A.
NATIONAL DEVELOPMENT SUPPORT COMMUNICATION
IN TEHERANIAN, M. ET AL (EDS) COMM POLICY FOR NATIONAL DEVELOPMENT
1977
P167-201
ROUTLEDGE & KEGAN PAUL: LONDON

TELECOMM INFO ECON SOC

687 PARKER, EDWIN B.
COMMUNICATION SATELLITES FOR RURAL DEVELOPMENT
1978, DEC
TELECOMM POLICY
2 N4 309-315

SAT INTERNAT TELECOMM RURAL
ECON

688 PARKER, EDWIN B.
COMMUNICATION SATELLITES FOR RURAL SERVICE (TELECOMM SYSTEMS)
1981, MARCH
TELECOMM POLICY (GB)
5 N1 12-17

SAT RURAL ECON

689 PARKER, EDWIN B.
COMMUNICATION SATELLITES FOR RURAL SERVICE PROBLEMS AND POTENTIALS
1981, MARCH
TELECOMMUNICATIONS POLICY
P12-17

TELECOMM SAT RURAL URBAN
TECHNICAL ECON POLICY

690 PARKER, EDWIN B.
COMMUNICATION TECHNOLOGY & SOCIAL POLICY: UNDERSTANDING THE NEW CULTURAL REVOLUTION. IN GERBNER, G., GROSS, L. & MELODY, W. (EDS)
1983
P533-545
JOHN WILEY: NEW YORK

TECH SOC ECON POLIT
POLICY CULTURE

691 PARKER, EDWIN B.
IMPLICATIONS OF NEW INFORMATION TECHNOLOGY
1973-74
PUBLIC OPINION QUARTERLY
37 P590-600

INFO TECH RES SOC
HIST ECON

692 PARKER, EDWIN B.
PLANNING COMMUNICATION TECHNOLOGIES AND INSTITUTIONS FOR DEVELOPMENT
1976, APR 5-10
PAPER/ EAST-WEST CONFERENCE ON COMMUNICATIONS POLICY AND PLANNING
STANFORD UNIVERSITY: PALO ALTO, CALIFORNIA

INFO TECH

693 PARKER, EDWIN B.& DUN, DONALD A.
INFORMATION TECHNOLOGY: ITS SOCIAL POTENTIAL
1972
SCIENCE
176

INFO TECH SOC EVAL

694 PARKER, WALTER B.
STATE OF ALASKA TELECOMMUNICATION POLICY: A KEY TO ALASKAN PARTICIPATION IN THE INFORMATION SOCIETY
1985
PARKER ASSOCIATES, INC.: ANCHORAGE, ALASKA

ALASKA TELECOMM

695 PARKER, WALTER B.
VILLAGE SAT II-THE SECOND EVALUATION OF SOME EDUCATIONAL USES OF THE ATS-1 SATELLITE EDUCATION BROADCASTING IN ALASKA
1973
IN TECH & EDUCATIONAL PROCEEDINGS OF NATIONAL CONFER ON TECH & ED

N AMER US ALASKA SAT
EDUC

696 PELTON, J.N. AND FILEP, R.T.
IN TOWARD INTERNATIONAL TELE-EDUCATION: TELE-EDUCATION VIA SATELLITE
BLUME, W.T. AND SCHNELLER, P.(EDS)
1984
P149-188
WESTERVIEW PRESS

EDUC INTERNAT SAT

697 PELTON, JOSEPH N.
DEVELOPMENT IN SATELLITE COMMUNICATIONS PRODUCES VALUABLE SPIN OFF FOR THE THIRD WORLD
1985
THIRD WORLD DEVELOPMENT
2 P90-95

SAT INTERNAT SOC ECON

698 PELTON, JOSEPH N.
COMMUNICATIONS DEVELOPING NATIONS FASTER
1984
SATELLITE COMMUNICATIONS
P19-22

SAT ECON

699 PELTON, JOSEPH N.
ECONOMIC AND POLICY PROBLEMS IN SATELLITE COMMUNICATIONS
IN SNOW, MARCELLUS, (ED)
1977
93-123
PRAEGER PUB: NY

POLICY INTERNAT SAT ECON
POLIT PLAN

700 PELTON, JOSEPH N.
GLOBAL COMMUNICATIONS SATELLITE POLICY: INTELSAT,
POLITICS AND FUNCTIONALISM
1974

SAT POLICY INTERNAT INTELSAT

701 PELTON, JOSEPH N.
GLOBAL THIN-ROUTE SERVICES USING INTELSAT: DIGITAL AND ANALOG
SOLUTIONS. IN TELECOMMUNICATIONS FOR PACIFIC DEVELOPMENT
1985
PAPER/ PACIFIC TELECOMMUNICATIONS CONFERENCE/ HONOLULU, HAWAII

INTELSAT TECHNICAL DIGITAL PAC

702 PELTON, JOSEPH N.
INTELSAT AND TELECOMMUNICATIONS DEVELOPMENT
1985, DEC 15-19
PAPER/JOINT AAS/JAPANESE ROCKET SOCIETY SYMPOSIUM/HONOLULU, HAWAII

INTELSAT TECHNICAL FIN BUSI
DIGITAL COST FIBER

703 PELTON, JOSEPH N.
INTELSAT, COMMUNICATIONS DEVELOPMENT AND WORLD COMMUNICATIONS YEAR
1985
1 N1 P75-85

WORLD INTELSAT

704 PELTON, JOSEPH N.
INTELSAT-THE GEOSYNCHRONOUS ORBITAL ARC AND THE THIRD WORLD COUNTRIES
1983
PAPER/ PACIFIC TELECOMMUNICATIONS COUNCIL/ HONOLULU, HAWAII
P136-145

INTELSAT ORBIT HIST COST
INVEST STAT EDUC

705 PELTON, JOSEPH N.
PEACEFUL USES OF OUTER SPACE: THE INTELSAT EXPERIENCE
1985, SEPTEMBER
INTERNATIONAL INSTITUTE OF COMMUNICATION: TOKYO

WORLD INTELSAT TELECOMM TECH
COST

706 PELTON, JOSEPH N.
RURALSAT: WHO HAS THE RIGHT ANSWER? AND WHEN?
1982, MARCH
SATELLITE COMMUNICATION
P29-34

SAT RURAL

707 PELTON, JOSEPH N.
THE COMMUNICATION SATELLITE: REVOLUTIONARY CHANGE AGENT?
1983
COLUMBIA J WORLD BUSINESS
18 SPR P77-84

SAT SOC ECON INTELSAT

708 PELTON, JOSEPH N. AND SNOW, MARCELLUS (EDS)
ECONOMIC AND POLICY PROBLEMS IN SATELLITE COMM
1977
PRAEGER PUBLISHERS: NEW YORK

INTERNAT TELECOMM MASS MEDIA US
POLICY

709 PELZ, D.C. & ANDREWS, F.M.
DETECTING CAUSAL PRIORITIES IN PANEL STUDY DATA
1964
AMER SOCIOLOGICAL REVIEW
29 P836-48

STAT RES

710 PERRATON, HILARY (ED.)
ALTERNATIVE ROUTES TO FORMAL EDUCATION: DISTANCE TEACHING FOR SCHOOL EQUIVALENCY
1982
THE JOHNS HOPKINS UNIVERSITY PRESS: BALTIMORE

RADIO EDUC RES BRAZIL
KOREA KENYA ISRAEL EVAL

711 PHUC, N.T. & DENNERY, G.
L'ECONOMIE DES TELECOMMS (COLLECTION SUP)
1972
PUF:PARIS

ECON TELECOMM

712 PIERCE W.B. AND JEQUIER, N.
CONTRIBUTION OF TELECOMMUNICATION TO THE ECONOMY
1977
TELECOMM J
44 N11 532-4

ITU ECON TELECOMM

713 PIERCE, W.B. AND JEQUIER, N.
TELECOMMUNICATIONS FOR DEVELOPMENT: SYNTHESIS REPORT OF THE ITU-OECD PROJECT ON CONTRIBUTION OF TELECOMM ....
1983
INTERNATIONAL TELECOMMUNICATION UNION: GENEVA

TELECOMM ECON SOC RES
EVAL ITU

714 PIERCE, W.B. AND JEQUIER, N.
TELECOMMUNICATIONS AND DEVELOPMENT
1979
DIALOGUE: WASHINGTON, D.C.
12 N2 P23

TELECOMM SOC ECON

715 PIERCE, JOHN R.
SIGNALS: THE TELEPHONE AND BEYOND
1981
W.H. FREEMAN: SAN FRANCISCO, CA

TELEPHONE TECH PLAN RES

716 PIERCE, WILLIAM B.
BENEFITS OF TELECOMMUNICATIONS TO DEVELOPMENT: INCLUDING FACTORS AFFECTING THE FINANCING OF RURAL TELECOMMUNICATIONS
1979, MAY28-JN9
SEMINAR ON RURAL TELECOMMUNICATIONS
HAMBURG, GERMANY

TELECOMM FIN RURAL REMOTE
COST TECH

717 PIERSON, J.H.S.
THE ECONOMIC NEED FOR TELECOMMUNICATIONS IN DEVELOPING COUNTRIES
1977
PAPER/ JOINT ORSA/TIME MEETING
ATLANTA, GEORGIA

TELECOMM ECON

718 PIETERSE, H.
MANAGEMENT OF TRAINING IN TELECOMMUNICATIONS
1980, MARCH
TELECOMMUNICATION JOURNAL
47 N3 P167-171

TELECOMM TRAIN MGT

719 PINET, A.
REFLECTIONS ON SOME DEVELOPMENT PROSPECTS FOR TELECOMMUNICATION NETWORKS
MARCH 1980
TELECOMM J. (ENG ED - SWITZERLAND)
47 N3 163-6

TECH TELECOMM

720 PINGLIER, A.
TELECOMMUNICATION SATELLITES FOR DEVELOPING COUNTRIES-EXTENSION OF SATELLITE USE TO RURAL
1984, FEB
ESA BULLETIN (FRANCE)
N21

TELECOMM SAT RURAL

721 PINHEIROR, G.
INVESTMENT AND TRAFFIC ASSIGNMENT IN A HYBRID GROUND AND SPATIAL COMMUNICATION NETWORK
1976
TECH REP #4

INVEST TECHNICAL TECH

722 PITRODA, SATAYAN G.
STATE OF TELECOMMUNICATIONS IN DEVELOPING COUNTRIES- AN OVERVIEW
1976, JULY
IEEE TRANSACTIONS ON COMMUNICATIONS
COM24 N7 P676-8

PAC TELECOMM INTELSAT RURAL
FIN

723 PITRODAS, D.
TELECOMMUNICATION DEVELOPMENT-THE THIRD WAY
1976
IEEE TRANSACTIONS ON COMMUNICATION
24 P736-42

TELECOMM

724 PLANT, CHRISTOPHER M.
PACIFIC ISLANDERS' USE OF PEACESAT: AN EVALUATION OF ITS APPROPRIATENESS FOR THE PACIFIC
P1-9
SIMON FRASER UNIVERSITY: VANCOUVER, CANADA

PAC SAT EDUC INFO

725 PLANT, CHRISTOPHER M.
TELECOMMUNICATIONS IN THE PACIFIC ISLANDS: HELPING OR HINDERING DEVELOPMENT?
P 19-27
SIMON FRASER UNIVERSITY: VANCOUVER, CANADA

PAC TELECOMM

726 PLANT, J., AMES, S
EMERGENCY-OUTPATIENT SATELLITES SERVE AS RURAL OUTPOSTS
1978
HOSPITALS - THE J OF THE AMERICAN HOSPITAL ASSOCIATION
52 N5 P87

RURAL SAT TECH HEALTH

727 PLUDE, FRANCIS FORDE
DIRECT BROADCAST SATELLITE IN AMERICA: DEFINING POLICY ISSUES THROUGH COOPERATIVE INTERACTION
1981
DISSERTATION ABSTRACTS INTERNATIONAL
45/06-A P1563

SAT BROADCAST SAT US

728 PODRACZKY, EMERIC
INTELSAT AND THE PACIFIC
1985, NOV 11-12
PAPER/12TH SOUTH PACIFIC REGIONAL TELECOMMUNICATION MEETING/KIRIBATI
SOUTH PACIFIC BUREAU FOR ECONOMIC COOPERATION: TARAWA, KIRIBATI

PAC INTELSAT TELECOMM HEALTH
EDUC SOC ECON FIN

729 POLCYN K.A.
THE ROLE OF COMMUNICATION SATELLITE IN EDUCATION AND TRAINING - THE 1990'S (EN)
1981
PROGRAMMED LEARNING AND EDUCATIONAL TECHNOLOGY
18 N4 P230-44

SAT EDUC TRAIN US
N AMER

730 POLCYN, KENNETH A.
AN EDUCATOR'S GUIDE TO COMMUNICATION SATELLITE TECHNOLOGY
1973
ACADEMY FOR EDUCAL DEVELOPMENT: WASHINGTON, D.C.

SAT TECH EDUC

731 POLISHUK, P. AND O'BRYANT, M.
TELECOMMUNICATIONS AND ECONOMIC DEVELOPMENT
1977
HORIZON HOUSE INTL: DEDHAM, MA

TELECOMM ECON INTERNAT SOC
EVAL RES

732 POLLARD, J.R.
TELECOMMUNICATIONS IN THE RURAL ENVIRONMENT
1983, OCT.
INFO. SCI.
2 N2 P107-21

N AMER TELECOMM RURAL

733 POOL, I.DE.S., STEVENS, P.M., KAMAL, A. AND DESSONKI, A.H.
APPROPRIATE TELECOMMUNICATIONS FOR RURAL DEVELOPMENT
1981, NOV 19
RESEARCH PROGRAM ON COMM POLICY AND TECH ADAPTATION PROGRAM
M.I.T. AND UNIVERSITY OF CAIRO

TELECOMM TECH RURAL

734 POOL, ITHIEL DE SOLA
FORECASTING THE TELEPHONE: A RETROSPECTIVE TECHNOLOGY ASSESSMENT
1983
ABLEX PUBLISHING: NORWOOD, NEW JERSEY

TELEPHONE EVAL

735 POOL, ITHIEL DE SOLA
INTERNATIONAL ASPECTS OF COMPUTER COMMUNICATION
1976, DEC
TELECOMM POLICY
1 N1 33-51

INTERNAT COMPUTER ECON POLICY
INFO DATA

736 POOL, ITHIEL DE SOLA
NEW STRUCTURES OF INTERNATIONAL COMMUNICATION? THE ROLE OF RESEARCH
MAIN PAPERS FROM THE 1980 CARACAS CONFERENCE IAMCR
1982
P60-75
LEICESTER, ADAM BROS & SHARDLOW

INTERNAT RES POLICY TECH
ECON POLIT

737 POOL, ITHIEL DE SOLA
SERVING THE INFORMATION NEEDS OF DEVELOPING COUNTRIES
CONFERENCE ON COMPUTER AND TELECOMMUNICATION POLICY
1976
INFORMATICS STUDIES N11

COMPUTER INTERNAT POLICY INFO

738 POOL, ITHIEL DE SOLA
THE RISE OF COMMUNICATION POLICY RESEARCH
1974, SPRING
JOURNAL OF COMMUNICATION
24 N2 P31-42

POLICY RES SOC ECON
POLIT TECH

739 POOL, ITHIEL DE SOLA (ED.)
THE SOCIAL IMPACT OF THE TELEPHONE
1977
MIT PRESS: CAMBRIDGE, MA

TELEPHONE SOC POLIT ECON

740 POOL, ITHIEL DE SOLA,
LOW COST DATA COMMUNICATION FOR LESS DEVELOPED COUNTRIES
1976
MIT PRESS: CAMBRIDGE, MA

DATA COST

741 POOL, ITHIEL DE SOLA, STONE, PHILIP AND SZALI, ALEXANDER (EDS)
COMMUNICATION, COMPUTERS & AUTOMATION FOR DEVELOPMENT
1971
UNITAR RES REPORT #6
UN INSTITUTE FOR TRAINING AND RESEARCH

COMPUTER TECH TELECOMM POLICY
INTERNAT

742 PORAT, MARC U.
GLOBAL IMPLICATIONS OF THE INFO SOCIETY
1978, WINTER
JOURNAL OF COMMUNICATION
28 N1 P70-80

INFO ECONOMY POLICY RES
US

743 PORAT, MARC U.
THE INFORMATION ECONOMY: DEFINITION AND MEASUREMENT
1977
US GOVERNMENT PRINTING OFFICE: WASHINGTON, D.C.

INFO ECON EVAL

744 PORTER, STEVE
AID STUDIES IN EDUCATIONAL TECHNOLOGY: PUBLIC SERVICE SATELLITES FOR DEVELOPMENT
1977, FEBRUARY
AID/CM/TA-147-548

AID TECH EDUC RES
SAT

745 POSTGATE, R., LEWIS, P.M., SOUTHWOOD, W.A.
LOW COST COMMUNICATIONS FOR EDUCATIONAL DEVELOPMENT PURPOSE IN THIRD WORLD COUNTRIES
1979, AUGUST
UNESCO PRESS: PARIS

EDUC COST

746 POSTGATE, RICHMOND, LEWIS, PETER AND SOUTHWOOD, WILLIAM (EDS)
APPROPRIATE COMMUNICATIONS IN THE THIRD WORLD.
1979, JULY
INTERMEDIA
7 N4 17-19

INTERNAT TECH POLICY

747 POWELL, LAWRENCE A.
TELECOMMUNICATION & THE ELDERLY: AUTONOMY OR SOCIAL CONTROL?
1985
ASSN PAPER-SSSP-SOCIETY FOR STUDY OF SOC PROBLEMS

TELECOMM SOC POLICY EVAL

748 PRASADA, BIRENDA
SOME IMPLICATIONS OF TELECOMMUNICATION POLICIES IN DEVELOPING COUNTRIES
1976
IEEE TRANSACTIONS ON COMMUNICATIONS
24 N7 P729-32

TELECOMM INDIA POLICY ECON

749 PRENTIS, M.R.
SOME ECONOMIC CONSIDERATIONS IN PLANNING TELECOMMUNICATION SERVICES DEVELOPMENT IN ISOLATED & UNDERPRIVILEGED AREAS OF CANADA
1978

CANADA TELECOMM PLAN ECON
RURAL REMOTE

750 PRIEST, M.J., CHAPMAN, S.A.
FINANCIAL AID FOR TELECOMMUNICATION IN DEVELOPING COUNTRIES
1979
TELECOMM J OF AUSTRALIA (AUSTRALIA)
29 N2

FIN AID TELECOMM

751 PRIEST, M.J., WHIPPY, J.F.D., GREENWOOD, P.
A FUTURE FOR SATELLITE COMMUNICATIONS IN PAPUA NEW GUINEA
1982
PACIFIC ISLANDS COMMUNICATION JOURNAL
2 N1 P23-28

PAPUA NEW GUINEA PAC SAT EDUC

752 PROBERT, D.
TOWARDS EXPERT SYSTEMS FOR TELECOMMUNICATION POLICY ANALYSIS
APRIL 1983
COMPUTER COMMUNICATION (GB)
6 N2 P58-64

POLICY TELECOMM UK

753 PURCELL, ARTHUR
SATELLITES AND DOMESTIC BROADCASTING
1985, JUNE/JULY
NATIONAL DEVELOPMENT ASIA
P65-69

SAT BROADCAST

754 PURTON, R.
RURAL TELECOMMUNICATION
1983, OCT
SYSTEM TECHNOLOGY (GB)
N37 P32-7

RURAL TELECOMM

755 RADA, JUAN F.
MICROELECTRONICS AND INFORMATION TECHNOLOGY: A CHALLENGE FOR RESEARCH IN THE SOCIAL SCIENCES
1980
SOC SCIENCE INFO
N2 435-465

INFO COMPUTER RES ECON

756 RADA, JUAN F.
THE IMPACT OF MICROELECTRONICS AND INFORMATION TECHNOLOGY: CASE STUDIES IN LATIN AMERICA
1982
UNESCO: PARIS

COMPUTER INFO ECON POLICY
L AMER

757 RADA, JUAN F.
THE MICROELECTRONICS REVOLUTION: IMPLICATIONS FOR THE THIRD WORLD
1981
DEVELOPMENT DIALOGUE
N2 41-67

INFO COMPUTER INTERNAT ECON
POLICY TECH

758 RADA, JUAN F.
THE SOCIO-ECONOMIC IMPACT OF MICROELECTRONICS.
IN BERTING, MILLS, WINTERSBERGER (EDS)
1980
P101-146
PERGAMON PRESS: OXFORD, NEW YORK

SOC ECON COMPUTER DIGITAL

759 RAGGET, R.J.
THE DEFT ANSWER FOR RURAL TELEPHONY
1982, DEC 20
TELEPHONY
P38-42

TELEPHONE RURAL ECON

760 RAGGET, R.J.
UP, UP AND AWAY IN NOVEMBER
1984, APRIL 23
TELEPHONY
P64,66

TELECOMM M EAST

761 RAGLAND, J. AND WARNER, P.
GREEN THUMB NATIONAL PILOT TEST
1981, FALL
EDUCOM BULLETIN
P16-19

US TELECOM RURAL COMPUTER
AGRIC V-TEXT

762 RAHIM, SYED A.
COMMUNICATION AND RURAL DEVELOPMENT IN BANGLADESH: CASE STUDY 3
1976
EAST-WEST COMMUNICATION INSTITUTE
EAST-WEST CENTER: HONOLULU, HAWAII

BANGLADESH RURAL CULT POLIT
HIST

763 RAMADASMURTHY, V., RAO, D.H., CLARENCE, I.D., AND BALASUBRAMANIAN,S.
NUTRITION EDUCATION AND SITE TELECASTS
1978
INTERNATIONAL JOURNAL OF HEALTH EDUCATION
21 N3 P168-173

SAT EDUC HEALTH

764 RAMAN, S.S.
ECONOMIC CONSIDERATION IN TELECOMMUNICATIONS DEVELOPMENT IN ISOLATED AND BACKWARD AREAS
1978
POST AND TELEGRAPH BOARDS: DELHI, INDIA

TELECOMM ECON REMOTE INDIA

765 RAMAN, S.S.,SANGAL D.K., PAI S.
INDIA'S RURAL TELECOMMUNICATION NETWORK
1979, DEC
TELECOMM (INDIA)
29 N2

INDIA RURAL TELECOMM TECH

766 RANDOLPH, MARC L. AND WADE, ROBERT E.
GLOBAL TELECOMMUNICATION: TRANSBORDER DATA FLOW AND THE FLOW OF BLACKS
1983-84 WINTER
URBAN LEAGUE REVIEW
8 N1
ERIC # EJ294300

TELECOMM POLICY INTERNAT INFO
SOC

767 RATNE, MALICCA
COMMUNICATION STRATEGY FOR LAND DEVELOPMENT AND SETTLEMENT
IN INTEGRATED COMM: BRINGING PEOPLE AND RURAL DEVELOPMENT TOGETHER
1975

RURAL PLAN

768 REED, WILLIAM E.
TELECOMMUNICATION: ECONOMIC STUDIES (BIBLIOGRAPHY WITH ABSTRACTS)
1979 & 78 JUNE
NTIS/PS-79/0603/5 & NTIS/PS-78/0581/5

TELECOMM ECON RES

769 REED, WILLIAM E.
TELECOMMUNICATION: ECONOMIC STUDIES.
(CITATIONS FROM THE NTIS DATA BASE)
1980 JUNE
PB80-811144 & PB80 811136

TELECOMM ECON RES

770 REYNOLDS, A., LAWLOR, E., CORRIGAN, J.A.O.
TELECOMMUNICATIONS
1980, MAY 22
BUSINESS AND FINANCE, DUBLIN
16 N36 P43

TELECOMM IRELAND

771 RICE, R.E., AND PARKER, E.B.
TELECOMMUNICATIONS ALTERNATIVES FOR DEVELOPING COUNTRIES
1979, AUTUMN
JOURNAL OF COMMUNICATION
24 N4 P125-136

TELECOMM SAT ECON COST

772 RIEGEL, O.W.
MASS MEDIA POLICIES AMD CHANGING CULTURES.
GERBNER, G.(ED)
1977
P63-72
JOHN WILEY & SONS: NY

INTERNAT POLICY MASS MEDIA CULTURE
SAT POLIT

773 RIZZONI, E.M., AND DEMENT, D.K.
IS THERE A SATELLITE IN KOREA'S FUTURE?
1984, MAY 21
TELEPHONY
P 35-48

KOREA SAT

774 RIZZONI, EITEL M.
AN OVERVIEW OF LATIN AMERICAN TELECOMMUNIATIONS, PAST
PRESENT AND FUTURE
1976
IEEE TRANSACTIONS ON COMMUNICATIONS
24 P 290-305

L AMER TELECOMM HIST EVAL

775 RIZZONI, EITEL M.
WORLDWIDE TELECOMMUNICATIONS: SYSTEM DEVELOPMENT PATTERNS
AND POLICIES
1977
IN TELECOMM AND ECONOMIC DEVELOPMENT, POLISHUK & O'BRYANT (EDS)

TELECOMM WORLD INTERNAT POLICY

776 ROBERT, F.
APPROPRIATE TECHNOLOGY IN LATIN AMERICAN TELECOMMUNICATION
IN POLISHUK & O'BRYANT, EDS, TELECOMM & ECON DEVELOPMENT
1977
HORIZON HOUSE INTL: DEDHAM, MA

L AMER TECH FIN INVEST

777 RODRIGUES, DIAS, ANTONIO, M., &; LEE, JOHN A.R.
NATIONAL COMMUNICATION POLICY COUNCILS: PRINCIPLES AND
EXPERIENCES
1979

MASS MEDIA POLICY INTERNAT BRAZIL
FINLAND

778 RODRIGUEZ, GABRIEL, ET AL.
L ERA TELEINFORMATICA: ARGENTINA
1985
FOLIOS EDICIONES ARGENTINA

ARGENTINA TELECOMM TECH

779 ROGERS, EVERETT M.
COMMUNICATION AND DEVELOPMENT: THE PASSING OF THE DOMINANT PARADIGM
1976
COMMUNICATION RESEARCH
3 P213-240

RES TELECOMM THEORY

780 ROGERS, EVERETT M.
COMMUNICATION STRATEGIES FOR AGRICULTURAL DEVELOPMENT
1972
ACADEMY FOR EDUCATIONAL DEVELOPMENT: WASHINGTON, D.C.

AGRIC TELECOMM

781 ROGERS, EVERETT M.
THE COMMUNICATION OF INNOVATIONS: COMBINING MEDIA AND INTER-PERSONAL CHANNELS
1972
IN INDUCING TECH CHANGE FOR ECON DEV. SOLO & ROGERS (EDS)

MODEL THEORY RES

782 ROGERS, EVERETT M.
DIFFUSION OF INNOVATIONS, THIRD EDITION
1983
FREE PRESS: NEW YORK

RES THEORY TELECOMM SOC
MEDIA

783 ROGERS, EVERETT M.
THE RISE AND FALL OF THE DOMINANT PARADIGM
1978, WINTER
JOURNAL OF COMMUNICATION
28, N1, P64-49

INTERNAT THEORY SOC CULT

784 ROGERS, EVERETT M. & SHOEMAKER. F.F.
COMMUNICATION OF INNOVATIONS
1971
FREE PRESS: NEW YORK

RES MEDIA THEORY

785 ROGERS, EVERETT M. AND ADHIKARYA, RONNY
COMMUNICATION AND INEQUITABLE DEVELOPMENT: NARROWING THE SOCIO-ECONOMIC BENEFITS GAP
1978
MEDIA ASIA
5 N1

SOC ECON

786 ROIZEN, J.
CHINA TURNS ON TV AND RADIO
1985
TELEVISION
22 N5 P251-255

CHINA TV RADIO CULT
SOC EDUC

787 ROSA, J.
RURAL TELECOMMUNICATION VIA SATELLITE
N 1981
TELECOMM
15 N11 P75-8

RURAL TELECOMM SAT

788 ROSENBERG, GERTRUDE
DISTANCE EDUCATION IN THE CANADIAN NORTH: AN ANNOTATED BIBLIO
OCCASIONAL PUB N12.
ED253135
ERIC DOC REPRO
CANADA

TELECOMM SAT RADIO EDUC
RURAL

789 ROSENBERG, VICTOR
NATIONAL INFORMATION POLICIES
1982, MAY 5
UNIVERSITY OF MICHIGAN: ANN ARBOR, MICHIGAN

TELEPHONE INFO POLICY RES

790 ROSENBERG, VICTOR
TELECOMMUNICATION DEVELOPMENT--THE THIRD WAY
1976, JULY
IEEE TRANSACTIONS ON COMMUNICATIONS
COM24N7 P736-42

TELECOMM TELEPHONE RES

791 ROSENDAAL, C. JANSEN VAN
AN INFORMATION AND TELECOMM POLICY FOR EUROPE
1983, AUG
BULLETIN OF THE AMERICAN SOCIETY FOR INFORMATION SCIENCE
9 N6 P7-11

ECON INFO POLICY EEC
INTERNAT TECH TELECOMM

792 ROWELL, D.M.
SOCIAL NEEDS AND TECHNOLOGY - A PARTNERSHIP FOR FUTURE
TELECOMMUNICATION DEVELOPMENTS IN AUSTRALIA
1980
TELECOMM J. AUST.
30 N2

AUSTRALIA PAC SAT SOC
TECH TELECOMM

793 RUSSELL, STEVEN
TECHNO-ECONOMICS OF US DOMESTIC SATELLITE ORBIT-SPECTRUM
UTILIZATION
1977
UNPUBLISHED DISSERTATION
STANFORD UNIVERSITY: PALO ALTO, CA.

ECON ORBIT SPECTURM US
SAT

794 RUTKOWSKI, ANTHONY M.
THE INTER-SATELLITE SERVICE - FUTURE SHOCK FOR TELECOMMUNICATION POLICY MAKERS
1981, JULY
TELECOMM
15 N7 P64-5

SAT TELECOMM POLICY

795 SACHDEV, DHARMENDRA K.
MICROWAVE SYSTEM DEVELOPMENT IN INDIA
1976, JULY
IEEE TRANSACTIONS ON COMMUNICATIONS
COM24 N7 P763

INDIA M-WAVE

796 SAKLOFSKE, D.H., TEATHER, D.C.B., ET AL.
CONTINUING EDUCATION AT A DISTANCE -- EVALUATING AUDIOVISUAL PACKAGES & TELEPHONE SEMINARS IN BEHAVIOR MODIFICATION ....
1977
PROGRAMMED LEARNING AND EDUC DEVELOPMENT
14 N1 13-25

TELEPHONE AV EDUC

797 SALINAS BASCUR, RAQUEL
TECHNOLOGY TRANSFER AND COMMUNICATION. IN HANCOCK, A. (ED)
1984
P123-167
PARIS:UNESCO

INTERNAT TECH POLICY ECON
POLIT CULT SAT L AMER
ASIA

798 SALINAS, RAQUEL
COMMUNICATION POLICIES: A CRUCIAL ISSUE FOR THE ACCOMPLISHMENT OF THE NEW INFO ORDER. THE CASE OF LATIN AMERICA
1978, MAY
RES PAPER SERIES #9

INTERNAT INFO POLICY L AMER

799 SANKAR, HOLLIS CARLYLE
SATELLITE TELECONFERENCING SERVING HIGHER EDUCATION: THE UNIVERSITY OF WEST INDIES EXPERIENCE
1984
THESIS, MASTERS

WEST INDIES CARIB SAT TELECONF
EDUC HIST TECH

800 SARDELLA, B. & DEGNAN, K.
SATELLITE BROADCASTING TO HOMES: SERVICE OPTIONS, ISSUES AND COSTS
1981, JUNE
TELECOMM POLICY
5 N2 P84-101

N AMER US SAT URBAN
POLICY ECON TV BROADCAST

801 SATHE, VASANT
HOW SATELLIES WILL EXTEND INDIA'S DEVELOPMENT COMMUNICATIONS
1981, SEPTEMBER
INTERMEDIA
9 N5 P37-40

INDIA SAT

802 SAUNDERS, R. AND DICKENSON, C.R.
TELECOMMUNICATIONS PRIORITY NEEDS FOR ECONOMIC DEVELOPMENT
1979
P1-10
WORLD BANK PUBLICATION: WASHINGTON, D.C.

TELECOMM ECON

803 SAUNDERS, R. AND WARFORD, J.
EVALUATION OF TELEPHONE PROJECTS IN LESS DEVELOPED COUNTRIES
(P.U. REPORT)
1978, JULY
P1-15
WORLD BANK PUBLICATION: WASHINGTON, D.C.

TELECOMM TELEPHONE FIN EVAL
INVEST REMOTE SOC

804 SAUNDERS, R. AND WARFORD, J.
TELECOMM PRICING AND INVESTMENT IN DEVELOPING COUNTRIES
IN POLISHUK & O'BRYANT, EDS, TELECOMM & ECON DEVELOPMENT
1977
HORIZON HOUSE INTL: DEDHAM, MA

TELEPHONE ECON FIN COST

805 SAUNDERS, R.H., AND WARFORD, J.
EVALUATION OF TELEPHONE PROJECTS IN LESS DEVELOPED COUNTRIES
1978
P.U. REPORT NO. PUN 37

WORLD TELEPHONE

806 SAUNDERS, R.H., WARFORD, J. AND WELLENIUS, B.
TELECOMMUNICATIONS AND ECONOMIC DEVELOPMENT
1983, JULY
(WORLD BANK) JOHNS HOPKINS UNIVERSITY PRESS: BALTIMORE, MD.

TELECOMM ECON FIN INVEST
REG MGT POLICY COST
URBAN RURAL

807 SAUNDERS, ROBERT
RURAL TELECOMMUNICATION: ECONOMIC AND POLICY IMPLICATIONS
1978
PAPER AT ITU SEMINAR ON RURAL TELECOMM, NEW DELHI

RURAL TELECOMM ECON POLICY

808 SAUNDERS, ROBERT
RURAL TELECOMMUNICATIONS ECONOMIC POLICY IMPLICATIONS
1978, SEPTEMBER
I.T.U. DOC #1-4, SEMINAR/ RURAL TELECOMMUNICATIONS
NEW DELHI, INDIA

TELECOMM RURAL ECON POLICY
ITU

809 SAUNDERS, ROBERT
TELECOMMUNICATION IN DEVELOPING COUNTRIES:
CONSTRAINTS ON DEVELOPMENT
1982
IN M. JUSSAWALLA & D.M. LAMBERTON (EDS) COMM ECON & DEV

WORLD TELECOMM ECON

810 SAUNDERS, ROBERT
TELECOMMUNICATIONS AND THE DEVELOPING WORLD: INVESTMENT DECISION AND PERFORMANCE MONITORING
1983, DECEMBER
TELECOMMUNICATION POLICY
7 N4 P277-284

TELECOMM INVEST FIN ECON
COST SOC

811 SCHENONE, OSVALDO & BAEZA, SERGIO
UN ANALISIS DE COSTOS Y BENEFICIOS DE TELEFONOS PUBLICOS
DE LARGA DISTANCIA EN ZONAS RURALES (ENGLISH SUMMARY)
1982, JN-AUG
ECONA
28

COST TELEPHONE RURAL

812 SCHILLER, HERBERT I.
INFORMATION AND CRISIS ECONOMY
(IN COMMUNICATION AND INFORMATION SCIENCE)
1984
ABLEX PUBLISHING: NORWOOD, NEW JERSEY

INFO ECON

813 SCHILLER, HERBERT I.
THE CRITICAL COMMUNICATIONS REVIEW, CHANGING PATTERNS OF
COMMUNICATIONS CONTROL IN MOSCO, V.& WASKO, J.
1984, 3-29
II
ABLEX PUB CORP: NORWOOD, NJ

THEORY RES SOC ECON
POLIT

814 SCHILLER, HERBERT I.
WHOSE NEW INTERNATIONAL ECONOMIC AND INFORMATION ORDER
1980
COMMUNICATION
5 P299-314

INTERNAT ECON POLIT THEORY

815 SCHNELLER, PAUL J.
FACILITATING GLOBAL EDUCATION WITH SATELLITE TELECOMMUNICATIONS
1982, MAY 19-22
PAPER/ CONFERENCE ON SHAPING THE FUTURE GLOBAL EDUCATION

WORLD SAT TELECOMM EDUC

816 SCHOLEY, D.H.A.
BRUNEI -- A TELECOMM BALANCING ACT
1981, JUNE 29
TELEPHONY
P95-96

BRUNEI TELECOMM TRAIN

817 SCHRAGE, G.S.
IS THERE A SATELLITE IN THE RURAL TELCO SKY?
1985 JUNE
TELEPHONY
206 N25

TELEPHONE RURAL SAT

818 SCHRAMM, W. & RUGGLES, W.W.
HOW MASS MEDIA SYSTEMS GROW
IN D. LERNER & W. SCHRAMM (EDS) COMM AND CHANGE IN DEV...
1967

MASS MEDIA

819 SCHRAMM, W., NELSON, L.M. AND BETHAN, M.T.
BOLD EXPERIMENT: THE STORY OF EDUCATIONAL TELEVISION IN AMERICAN SAMOA
1981
STANFORD UNIVERSITY PRESS: PALO ALTO, CA

AMERICAN SAMOA TV EDUC EVAL

820 SCHRAMM, WILBUR
BIG MEDIA, LITTLE MEDIA
1977
SAGE PUBLISHING: BEVERLEY HILLS, CA.

MASS MEDIA TECH

821 SCHRAMM, WILBUR
MASS MEDIA AND NATIONAL DEVELOPMENT
1964
STANFORD UNIVERSITY PRESS: PALO ALTO, CA.

MEDIA POLIT ECON EVAL

822 SCHRAMM, WILBUR & LERNER, DANIEL (EDS)
COMMUNICATION AND CHANGE: THE LAST TEN YEARS - AND THE NEXT
1976
UNIVERSITY PRESS OF HAWAII: HONOLULU, HAWAII

INTERNAT MASS MEDIA THEORY RES
SOC CULT

823 SCHURINGA, T.M.
PLANNING OF TELECOMMUNICATION SERVICES IN RURAL AREAS IN NON-INDUSTRIALISED COUNTRIES
1983, JUNE
PHILIPS TELECOMM. REV. (NETHERLANDS)
41 N2 P100-12

WORLD TELECOMM RURAL PLAN

824 SCHURINGA, T.M.
THE MICROCHIP IN THE CRYSTAL BALL- 25 YEARS ON...
1983, JAN/FEB
DEVELOPMENT FORUM
P8

COMPUTER

825 SCHWARTS, LLOYD
SAY U.S. LAGS IN BID TO TAP CHINA TELECOMMUNICATION MARKETING
1985, MAY 27
ELECTRONIC NEWS
P33

US CHINA TELECOMM

826 SCRAFFORD, ROBERT L.
A COMMUNICATIONS EXPLOSION IN OMAN
1976, JULY
IEEE TRANSACTIONS ON COMMUNICATIONS
COM24 N7 P691

OMAN TELECOMM INTERNAT

827 SEGAL, B.
ITU PLENIPOTENTIARY CONFERENCE AND BEYOND. A CASE FOR SERIOUS FOREIGN POLICY (TELECOMM)
1983, DEC
TELECOMM POLICY (GB)
7 N4 P326-34

ITU TELECOMM INTERNAT POLICY

828 SEGOVIA, A.
IMPLEMENTATION OF SPACE SATELLITE REMOTE SENSING PROGRAMS IN DEVELOPING COUNTRIES (ECUADOR)
1982, MAY
N83-19189/0

ECUADOR SAT DATA REMOTE
MODEL

829 SERRANO, ARTURO, AND VALERDI, JORGE
MANUFACTURING STRATEGIES FOR LOW-CAPACITY EARTH STATIONS IN DEVELOPING COUNTRIES
1979, SEP 24-26
PAPER/ THIRD WORLD TELECOMMUNICATIONS FORUM/ GENEVA, SWITZERLAND

TELECOMM SAT PLAN

830 SERVAN-SCHREIBER, JEAN-JACQUES
THE WORLD CHALLENGE
1981
SIMON AND SCHUSTER: NEW YORK

WORLD TECH SOC ECON

831 SESAY, S.J.
SOME PROBLEMS OF IMPROVING TELECOMM SYSTEMS IN THE DEVELOPING COUNTRIES
IN POLISHUK & O'BRYANT, EDS, TELECOMM & ECON DEVELOPMENT
1977
HORIZON HOUSE INTL: DEDHAM, MA

INTERNAT TELECOMM PLAN AFRICA

832 SETZER, F.O., FRANCA, B.A., CORNELL, N.W.
POLICIES FOR REGULATION OF DIRECT BROADCAST SATELLITES
1980, SEPT
PB81-151201

SAT POLICY REG BROADCAST

833 SHACKMAN, A.D.
PRIVATE BUSINESS COMMUNICATION IN DEVELOPING COUNTRIES
1984
COMM SAT SYSTEMS CONF 10TH ORLANDO FL TECH PAPERS
A84-2525110-32
AMER INST OF AERONAUTICS & ASTRONAUTICS NY

BUSI TELECOMM ECON

834 SHADAT, S.A.
FARM BROADCASTING IN RADIO BANGLADESH
1982, DECEMBER
COMBROAD
P15-17

BANGLADESH RADIO AGRIC

835 SHANTA, B.H.
POWER SOURCES FOR RURAL TELECOMM
DEC. 1979
TELECOMM (INDIA)
29 N2 P201-4

INDIA RURAL TELECOMM TECH

836 SHANTA, B.H.
TECHNOLOGY OF SWITICHING FOR RURAL TELECOMMUNICATION
DEC. 1979
TELECOMM (INDIA)
29 N2

TELEPHONE TELECOMM RURAL TECH
INDIA

837 SHAPIRO, P.S.
COMMUNICATION OR TRANSPORT: DECISION-MAKING IN DEVELOPING
COUNTRIES
1967
MIT CENTER FOR INTERNATIONAL STUDIES: CAMBRIDGE, MA.

WORLD TRANSP

838 SHAPIRO, P.S.
TELECOMMUNICATION AND INDUSTRIAL DEVELOPMENT
1976
IEEE TRANSACTIONS ON COMM
24 P305-311

TELECOMM BUSI ECON

839 SHARMA, R.G.
RURAL COMMUNICATIONS PLANNING METHODOLOGY FOR INTEGRATING
SATELLITE AND TERRESTRIAL FACILITIES
1976
TECH REPORT 66
COMM SATELLITE PLANNING CENTER, STANFORD UNIV: STANFORD, CA.

RURAL SAT PLAN

840 SHARMA, R.G.
RURAL COMMUNICATIONS PLANNING METHODOLOGY FOR INTEGRATING
SATELLITE AND TERRESTRIAL FACILITIES
1977
DISSERTATION
UNIVERSITY MICROFILMS INTERNATIONAL: ANN ARBOR, MI

SAT RURAL PLAN

841 SHARP, ERIC
TELECOMMUNICATION AND ECONOMICS
1985, MAY 6
TELEPHONY
20 N18 P197-202

TELECOMM ECON

842 SIEBERT, D.J.
A DECADE OF EXPERIENCE USING TWO-WAY CLOSED CIRCUIT TELEVISION
FOR MEDICAL CARE AND EDUCATION
1977, NOVEMBER
REPORT/LISTER HILL NATIONAL CENTER FOR BIOMEDICAL COMMUNICATIONS
NATIONAL LIBRARY OF MEDICINE: BETHESDA, MARYLAND

TV VIDEO HEALTH EDUC

843 SIGNITZER, BENNO
REGULATION OF DIRECT BROADCASTING FROM SATELLITES: THE
UN INVOLVEMENT
1976

INTERNAT SAT UN POLICY
REG

844 SILBERMANN, ALPHONSE
PROBLEMS OF SCIENTIFIC INVESTIGATION OF SOCIAL ASPECTS OF SATELLITE BROADCASTING
1980
COMM

SOC SAT BROADCAST RADIO
TV

845 SILCOX, A.R., DANDATO, M.F.,DIXON, A.
TELECOMMUNICATION & POSTAL DEVELOPMENT PLANS FOR ZIMBABWE
1983, SEPT
ZIMBABWE ENG
21 N5

ZIMBABWE TELECOMM PLAN POLIT
TELEPHONE

846 SIMPSON, A. A.
REMOTE COMMUNICATION IN CANADA AND SOCIAL INTERACTION--A MIDDLE MANAGEMENT PERSPECTIVE
1978
CANADIAN DEPT OF COMMUNICATIONS: OTTAWA

CANADA SOC MGT

847 SINCLAIR, JAMES
UNITING A NATION: THE POSTAL AND TELECOMMUNICATION SERVICE OF PAPUA NEW GUINEA
1984
OXFORD UNIVERSITY PRESS: MELBOURNE, AUSTRALIA

PAPUA NEW GUINEA PAC TELECOMM HISTORY

848 SINGER, BENJAMIN D.
SOCIAL FUNCTIONS OF THE TELEPHONE
1981
R & E RESEARCH ASSOCIATES: PALO ALTO, CA.

SOC TELEPHONE

849 SINGH, B. INDU (ED.)
TELECOMMUNICATIONS IN THE YEAR 2000: NATIONAL AND INTERNATIONAL PERSPECTIVE
1983
ABLEX PUBLISHING: NORWOOD, NJ

TELECOMM US WORLD INFO
TECH POLICY

850 SINGH, K.J. AND GROSS, B.M.
THE CHALLENGES OF COMMUNICATION FOR RURAL RECONSTRUCTION
1979
MEDIA ASIA
P2-14

RURAL

851 SINHA, P.R.R. (ED)
COMMUNICATION AND RURAL CHANGE
1976
ASIAN MASS COMMUNICATION RESEARCH AND INFORMATION CENTRE: SINGAPORE

RURAL

852 SITES, M.
A DEMAND ASSIGNMENT MULTIPLE ACCESS (DAMA) CONTROL SYSTEM FOR THIN-ROUTE SAT TELEPHONE SYSTEMS
1976
TECH REPT # 10

SAT TELEPHONE DATA

853 SITES, MICHAEL
A DEMAND ASSIGNMENT MULTIPLE ACCESS (DAMA) CONTROL SYSTEM FOR THIN-ROUTE SATELLITE TELEPHONE SYSTEMS
1976
COMM SATELLITE PLANNING CENTER, STANFORD UNIV: STANFORD, CA.

SAT TELEPHONE RURAL TECH

854 SLOTANI-SHIRAZI, ALIREZA
COMMUNICATION SATELLITES IN THE 80'S: TECHNO-ECONOMIC FACTORS OF DELIVERY SYSTEMS FOR ETV/ITV IN LESS-DEVELOPED REGIONS
1979
DISSERTATION ABSTRACTS INTERNAT
40/08-A P4366

SAT TECH ECON TV
EDUC

855 SMITH, ANTHONY
THE GEOPOLITICS OF INFORMATION: HOW WESTERN CULTURE DOMINATES THE WORLD
1980
OXFORD UNIVERSITY PRESS: OXFORD, NEW YORK

INTERNAT SOC POLIT TECH
SAT

856 SMITH, DELBERT D.
CHINA SATELLITE COMMUNICATION
1980, JANUARY
SATELLITE COMMUNICATIONS
P20-25

CHINA SAT

857 SMITH, DELBERT D.
COMMUNICATION VIA SAT: A VISION IN RETROSPECT
1976
A.W. SIJTHOFF: LEYDEN, BOSTON

SAT EVAL HIST

858 SMITH, DELBERT D.
THE IMPACT OF SPACE POLICY ON DIRECT BROADCAST SATELLITES
MAY 1980
SAT COMM
4 N5

SAT POLICY NASA

859 SMITH, W.A., PAREJA, R., BOOTH, E., AND KASMUSON, M.
HEALTH COMMUNICATION FOR ORT IN HONDURAS
1984
ASSIGNMENT CHILDREN
65-68
UNICEF: GENEVA

HONDURAS RADIO MEDIA EDUC
TRAIN RURAL HEALTH

860 SNOW, MARCELLUS S.
INTERNATIONAL COMMERCIAL SATELLITE COMMUNICATIONS: ECONOMIC & POLITICAL ISSUES OF THE FIRST DECADE OF INTELSAT
1976, JAN
SERIES: SPECIAL STUDIES

INTERNAT SAT ECON POLIT
INTELSAT

861 SOEWANDI, K. AND SOEDARMADI, P.
TELECOMMUNICATIONS IN INDONESIA
1976, JULY
IEEE TRANSACTIONS ON COMMUNICATIONS
COM24 N7 P687

INDONESIA TELECOMM

862 SOROOS, MARVIN S.
THE COMMONS IN THE SKY: THE RADIO SPECTRUM & GEOSYNCHRONOUS ORBITS AS ISSUES IN GLOBAL POLICY
1982, SUMMER
INTERNATIONAL ORG
36 P665-77

WORLD SAT SPECT ORBIT

863 SOULE, LARRY
WHY THE SUDAN NEEDS A TWO-WAY RADIO
1980, JULY
INTERMEDIA
8 N4 P12-16

SUDAN 2-WAY RADIO RURAL

864 SOULE, LARRY AND VIVIAN
SPREADING THE SCIENTIFIC WORD IN AFRICA
1985, MARCH
INTERMEDIA
13 N12 P35-39

AF

865 SOUTHWORTH, J.H., FLANIGAN, J.M., AND KNEZEK, J.A.
COMPUTERS IN EDUCATION: INTERNATIONAL MULTIMODE ELECTRONIC CONFERENCING
1980, JULY
PAPER/ EAST-WEST COMMUNICATION INSTITUTE: HONOLULU, HI

PAC COMPUTER EDUC REMOTE

866 SPAIN, PETER L., JAMISON, DEAN T., AND MCANANY EMILE G.,(EDS)
RADIO FOR EDUCATION AND DEVELOPMENT CASE STUDIES
1977
WORLD BANK: WASHINGTON, D.C.

RADIO EDUC

867 SPARKES, VERNONE M.
THE PUBLIC COMMUNICATIONS PLANNER: RESPONDING TO THE NEW TELECOMMUNICATIONS OPPORTUNITIES
1976
PAPER AT ADVANCE SUMMER SEMINAR ON COMM TECH

TELECOMM PLAN

868 SPERO, J.
INFORMATION AND TELECOMMUNICATION AS A TRADE ISSUE
1982, MARCH
INTERMEDIA
10 N2 P9-11

INFO TELECOMM INTERNAT ECON
POLICY

869 STEEVES, W. DEWITT, J
SUCCESSFUL TRANSFER OF US TELECOMMUNICATION & RELATED TECHNOLOGY TO THIRD WORLD; CORE COMPETENCIES FOR INDUSTRY MANAGERS
1984
DISS. ABSTRACTS INTERNAT
45/08-A P2422

TELECOMM INTERNAT US TECH
MGT

870 STERLING, CHRISTOPHER H.
TRACKING THE NEW TECH. A SUMMARY OF THE SECOND INVITATIONAL POSTSECONDARY EDUCATION REVIEW PANEL
ED247912
ERIC DOC REPRO SERV

TELECOMM SAT EDUC

871 STERLING, CHRISTOPHER H.(ED)
INTERNATIONAL TELECOMMUNICATION & INFORMATION POLICY
1984
COMMUNICATION PR INC.: WASHINGTON, D.C.

TELECOMM INTERNAT INFO POLICY

872 STEWART, FRANCES, & COOPER, C. (EDS)
SCIENCE, TECHNOLOGY AND DEVELOPMENT. THE POLITICAL ECONOMY OF TECHNICAL ADVANCE IN UNDERDEVELOPED COUNTRIES
1977
P99-121
FRANK CASS & CO: LONDON

TECH ECON POLIT INTERNAT

873 STREET, JAMES AND DILMUS, JAMES (EDS)
TECHNOLOGICAL PROGRESS IN LATIN AMERICA: THE PROSPECTS FOR OVERCOMING DEPENDENCY
1979
WESTVIEW PRESS: BOULDER, CO.

TECH ECON RES L AMER

874 STUART, SARA
VIDEO IN THE VILLAGE
1986, SPRING
DEVELOPMENT COMMUNICATION REPORT
P7-8

VIDEO RURAL

875 STUART, SARA
VILLAGE SOLUTIONS MAKE GLOBAL COMMUNITY
1983, FEB 23-24
DEVELOPMENT FORUM
UNITED NATIONS: SWITZERLAND

WORLD RURAL

876 SUBRAMANIAN, T.S.
TELECOMMUNICATION IN AID OF NATIONAL DEVELOPMENT (IN INDIA)
1983, DEC
TELECOMM (INDIA)
33 N2 P3-7

TELECOMM INDIA ECON HIST

877 SUCHMAN, D., AUVINE, B.A., HINTON, B.H.
A STUDY OF THE ECONOMIC BENEFITS OF METEOROLOGICAL SATELLITE DATA
1980, SEPT
N81-25631/5

SAT DATA ECON EVAL

878 SUSSMAN, G.
GLOBAL TELECOMMUNICATIONS AND THE THIRD WORLD: THEORETICAL CONSIDERATIONS
1984, JULY
MEDIA, CULTURE AND SOCIETY
6 N3 P289-800

WORLD TELECOMM ECON SOC
POLIT CULT

879 SUSSMAN, G.
TELECOMMUNICATIONS TECHNOLOGY: TRANSNATIONALIZING THE NEW PHILIPPINE INFORMATION ORDER
1982, OCT
MEDIA, CULTURE AND SOCIETY
4 N4 P377-390

PHILIPPINES TELECOMM COMPUTER POLIT
ECON FIN

880 SWAMINATHAN, K.
RESEARCH AND DEVELOPMENT IN TELECOMMUNICATIONS IN DEVELOPING COUNTRIES
1983
TELECOMMUNICATION JOURNAL
50 N3 P140-146

TELECOMM RES INTERNAT TECH

881 SWARTZ, M.R., CHAAD D.C., EVENS, F.W., & DOHNER C.W.
SATELLITE IN HEALTH EDUCATION AND HEALTH CARE PROVISION - THE WAMI EXPERIENCE
1983
J OF THE AMER MEDICAL ASSOC
250 N5 636-639

SAT HEALTH TELEMEDICINE EDUC
WAMI RURAL

882 SWEENEY, G.P.
TELEMATICS AND DEVELOPMENT
1981
THE INFORMATION SOCIETY
1 N2 P113-132

TELECOMM INFO COMPUTER ECON

883 SY, DEMBA JACQUES HABIB
CAPITALIST MODE OF COMMUNICATION, TELECOMM UNDERDEVELOPMENT & SELF-RELIANCE APPROACH TO TELECOMM HISTORY AND SAT PLAN IN AF
1984
DISSERTATION ABSTRACTS INTERNATIONAL
45/12-A P3474

AF TELECOMM SAT HIST
ECON

884 SYZMANSKI, RAY
TREMPEALEAU COUNTY KELLOGG PROJECT: FINAL PROJECT SUMMARY
1981
PENN STATE UNIVERSITY: STATE COLLEGE, PA.

EDUC CABLE TV SAT
TELECOMM

885 TAGASAKI, NOZOME, ET.AL.
A STUDY ON PACIFIC REGIONAL SATELLITE COMMUNICATIONS
1983
PAPER/ PACIFIC TELECOMMUNICATION CONFERENCE '83/ JAPAN

PAC SAT

886 TANHUNAA, A. R.
THE SOCIO-ECONOMIC IMPORTANCE OF TELETRAFFIC DEVELOPMENT
1974
SAHKO:ELECTRICITY IN FINLAND
47 (5-6)

TELECOMM SOC ECON

887 TASKAR, N.T. AND RAMAKRISHNAN, S.
INDIA: A LOOK AT TELECOMMUNICATION PROGRESS, DEVELOPMENT ON THE SUB-CONTINENT
1981, MAY 25
TELEPHONY
200 N21

INDIA TELECOMM TECH ECON
SOC SAT

888 TAWNEY J.W., AESCHLEMAN, S.R., DEATON, S.L., DONALDSON, R.M.
USING TELECOMMUNICATION TECHNOLOGY TO INSTRUCT RURAL SEVERELY HANDICAPPED CHILDREN
1979
EXCEPTIONAL CHILDREN
46 N2 P118-25

TELECOMM RURAL EDUC TECH
HEALTH

889 TAYLOR, DAVID
DIAL THE FUTURE
1985
THIRD WORLD DEVELOPMENT
2 P194-195

TELECOMM

890 TAYLOR, RICHARD DENNY
SATELLITE DIRECT BROADCAST: THE PROSPECTS FOR DEVELOPMENT
1978
DISSERTATION ABSTRACTS INTERNAT
39/01-A P11

SAT ECON

891 TEDROS, G.
THE DEVELOPMENT OF THE PAN-AFRICAN TELECOMMUNICATION
NETWORK (PANAFTEL 1960-1983)
1984, APRIL
TELECOMM J
51 N4 P204-10

PANAFTEL HIST TELECOMM AF

892 TEER, K.
COMMUNICATION AND TELECOMMUNICATION
IEEE TRANSACTIONS ON COMMUNICATIONS
23 P1040-45

TELECOMM THEORY RES

893 TEHRANIAN, MAJID
THE COURSE OF MODERNITY: THE DIALECTICS OF MODERNIZATION
AND COMMUNICATION
1980
INTERNAT SOC SCIENCE J
32 P247-262

THEORY INTERNAT TECH POLIT

894 TEHRANIAN, MAJID
THE FETISH OF IDENTITY: COMM REVOLUTIONS AND FUNDAMENTAL REVIVALS
1981
MEDIA ASIA
8 N1 P35-39

TELECOMM POLIT IRAN

895 TEHRANIAN, MAJID, ET AL
COMMUNICATIONS POLICY FOR NATIONAL DEVELOPMENT
1977
ROUTLEDGE AND KEGAN PAUL: BOSTON

POLICY POLIT SOC ECON

896 TELL, BJORN V.
THE AWAKENING INFORMATION NEEDS OF THE DEVELOPING COUNTRIES
1980
JOURNAL OF INFORMATION SCIENCE
P285-289

INFO MODEL

897 TENOPIR, C. AND JACKSON, M.M.
TELECOMMUNICATION AND PUBLISHING IN THE SOUTH PACIFIC
1984, SEPTEMBER
ELECTRONIC PUBLISHING REVIEW
4 N3 P189-200

SOUTH PAC SAT HIST INFO
SOC MEDIA TELEX

898 TEUBAL, M. AND STEINMUELLER, E.
GOVERNMENT POLICY INNOVATION AND ECONOMIC GROWTH: LESSONS FROM A STUDY OF SATELLITE COMMUNICATIONS
1982, OCT
RESEARCH POLICY
11 N5 P271-287

POLICY SAT HIST TECH
ECON NASA

899 TEXTOR, ROBERT B. ET AL
ANTICIPATORY ANTHROPOLOGY AND THE TELEMICROELECTRONIC REVOLUTION: A PRELIMINARY REPORT FROM SILICON VALLEY
1985, SPRING
ANTHROPOLOGY & EDUC QUARTERLY
16 N1 P30

US TELECOMM COMPUTER SOC POLICY

900 THORNGREN, B.
SILENT ACTORS: COMMUNICATIONS NETWORKS FOR DEVELOPMENT
1977
IN THE SOCIAL IMPACT OF THE TELEPHONE, I. POOL (ED)
MIT PRESS: CAMBRIDGE, MA.

TELECOMM ECON EVAL

901 TIGRE, BASTOS
TECHNOLOGY AND COMPETITION IN THE BRAZILIAN COMPUTER INDUSTRY
1983
ST. MARTIN'S PRESS: NEW YORK

COMPUTER TECH POLICY ECON BRAZIL

902 TIMMERS, A.G.W.
SATELLITE SYSTEMS FOR RURAL DEVELOPMENT IN AFRICA
1981, MARCH
REGIONAL SEM ON REMOTE SENSING APPLIC AND SAT COMM FOR EDUC & DEV

AF SAT RURAL ETHIOPIA

903 TOURZAN, REZA G.A.
IRAN'S PRESENT TELECOMMUNICATION SYSTEM AND ITS EXPECTED DEVELOPMENT
1976, JULY
IEEE TRANSACTIONS ON COMMUNICATIONS
COM24 N7 P694

IRAN TELECOMM TV EDUC

904 TRAUBENBERG, I. A. AND PATRUNOVA, Z.E.
ON THE INFLUENCE OF THE MEANS OF COMMUNICATION
ON THE PROCESS OF MANAGEMENT
1973
TRUDY UCEBNYCH INSTITUTVSHJAZI (IN RUSSIAN)
65

USSR MGT

905 TRUEBLOOD, CECIL R. AND FLANAGAN, KATHLEEN R.
THE ROLE OF MICROCOMPUTERS IN RURAL SCHOOLS AND TECHNOLOGY AND EDUCATION
PROCEEDINGS OF NATIONAL CONFERENCE ON TECHNOLOGY AND EDUCATION

TELECONF RURAL COMPUTER EDUC
PENNSYLVANIA

906 TYLER, M.
THE IMPACT OF TELECOMMUNICATIONS ON THE PERFORMANCE OF A SAMPLE OF BUSINESS ENTERPRISES IN KENYA
1981
INTERNATIONAL TELECOMMUNICATION UNION: GENEVA

AF KENYA TELECOMM BUSI

907 TYLER, M., MORGAN, R., CLARKE, A.
TELECOMMUNICATIONS AND ENERGY POLICY
1981
INTERNATIONAL TELECOMMUNICATION UNION: GENEVA

TELECOMM POLICY

908 TYLER, MICHAEL
ECONOMIC EVALUATION OF TELECOMMUNICATIONS INVESTMENTS IN DEVELOPING COUNTRIES
1983, DECEMBER
PAPER/ WORLD COMMUNICATION YEAR SEMINAR IN KUALA LUMPUR, MALAYSIA

ECON EVAL TELECOMM INVEST

909 TYLER, MICHAEL
ENHANCED COMMUNICATION SERVICES FOR DEVELOPING COUNTRIES
(COMMUNICATIONS STUDIES AND PLANNING INTERNATIONAL INC. (CSPI), N.Y.
1983
PROCEEDINGS OF THE WORLD TELECOMM FORUM, INTL TELECOMM UNION
INTERNATIONAL TELECOMMUNICATION UNION: GENEVA

TELECOMM TELEPHONE TECH T-TEXT
DATA

910 TYLER, MICHAEL
THE EVALUATION OF TELECOMM PROGRAMS AND PROJECTS IN LESS DEVELOPED
COUNTRIES (IN POLISHUK AND O'BRYANT) EXPOSITION PROCEEDINGS VOL 1
1978
PAPERS/ INTELCOM`77/ ATLANTA, GEORGIA
P93-129
HORIZON HOUSE INTERNATIONAL: DEDHAM, MA.

TELECOMM EVAL ECON INTERNAT
POLICY

911 UDOFIA, CALLIX IME
AFRICA & THE INTERNATIONAL TELECOMMUNICATION UNION
1981
DISSERTATION ABSTRACTS INTERNAT
42/11-A P4636

AF ITU

912 UME-NWAGBO, EBELE N. E.
BROADCASTING IN NIGERIA: ITS POST-INDEPENDENCE STATUS
1984, AUT
JOURNALISM QUARTERLY
16 N3 P585-92

NIGERIA MEDIA RADIO TV
ECON

913 UNITED NATIONS
STATISTICAL YEARBOOK
ANNUAL
UNITED NATIONS: NEW YORK

STAT UN

914 UNITED NATIONS
UN TRANSPORTATION AND COMMUNICATION DECADE FOR AFRICA
1978-1988
1979
I-III
UN: ADDIS ABABA
PANAFTEL

RURAL SAT BROADCAST FIN
AF UN TRANS

915 UNITED NATIONS CENTER FOR TRANSNATIONAL CORPORATIONS
TRANSNATIONAL CORPORATIONS AND TRANSBORDER DATA FLOWS:
A TECHNICAL PAPER
1982
ST/CTC/23
UNITED NATIONS: NY

INTERNAT DATA COMPUTER ECON
POLICY UN

916 UNITED NATIONS/UNESCO
INTERGOVERNMENTAL CONFER ON COMMUNICATION POLICIES IN LATIN AMERICA
AND THE CARIBBEAN
1976
FINAL REPORT

TELECOMM MASS MEDIA POLICY L AMER
CARIB UNESCO

917 UNITED NATIONS/UNESCO
INTERGOVERNMENTAL CONFERENCE ON COMMUNICATION POLICIES IN AFRICA
1980
FINAL REPORT
UNESCO: PARIS

TELECOMM MASS MEDIA POLICY AF
UNESCO

918 UNITED NATIONS/UNESCO
INTERGOVERNMENTAL CONFERENCE ON COMMUNICATION POLICIES IN ASIA AND
OCEANIA
1979
FINAL REPORT
UNESCO: PARIS

TELECOMM MASS MEDIA POLICY
L AMER ASIA PAC UNESCO

919 UNITED NATIONS/UNESCO
INTERGOVERNMENTAL COUNCIL OF THE INTERNATIONAL PROGRAMME
FOR THE DEVELOPMENT OF COMMUNICATION (SECOND SESSION)
1982
FINAL REPORT

TELECOMM INTERNAT PLAN UNESCO

920 UNESCO
REVISED GUIDELINES FOR ECONOMIC EVALUATION OF NATIONAL
COMMUNICATION SYSTEMS
1974
PAPER/ UNESCO SEMINAR ON TRAINING OF COMM PLANNERS
UNESCO: PARIS

ECON EVAL

921 UNITED NATIONS/UNESCO
STATISTICAL YEARBOOK
ANNUAL
UNESCO

STAT UNESCO

922 UNITED NATIONS/UNESCO
TELECOMMUNICATIONS TARRIFFS FOR THE MASS MEDIA
NEW COMM ORDER 8
UNESCO: PARIS

TELECOMM ECON REG INTERNAT
UNESCO

923 US SENATE COMMITTEE ON COMMERCE, SCIENCE & TRANSPORTATION SUBCOMM
RURAL TELECOMM HEARING
1981, JUNE 22
IN TECHNOLOGY & EDUCATION PROCEEDINGS OF NAT CONF ON TECH & ED
SERIAL 97-60
97TH CONGRESS, 1ST SESSION

N AMER US TELECOMM RURAL
POLICY

924 US SENATE COMMITTEE ON COMMERCE, SCIENCE, AND TRANSPORTATION
COMMUNICATIONS AND RURAL AMERICA
1978
SERIES: 95TH CONGRESS, 1ST SESSION
DOC. TYPE: M

RURAL CABLE TV SAT
US

925 VALASKAKIS, GAIL
COMMUNICATION AND THE CANADIAN NORTH
1983
P120-135
CONCORDIA UNIV, DEPT OF COMM STUDIES: MONTREAL

RURAL SAT TV SOC
CULT CANADA

926 UNESCO
LIST OF DOCUMENTS AND PUBLICATIONS IN THE FIELD OF
MASS COMMUNICATION (1984-1985)
1986
UNESCO: PARIS

RADIO TV TELECOMM SAT
SOC CULT RES PLAN

927 VALERDI, JORGE
A COMMUNICATIONS POLICY FOR MEXICO
1977, SEPTEMBER
TELECOMMUNICATIONS POLICY
1 N4 P271-288

MEXICO TELECOMM ECON POLICY
TELEPHONE TELEX TV

928 VALERDI, JORGE
MEXICAN TELECOMMUNICATIONS FOR THE YEAR 2000
IN POLISHUK & O'BRYANT, EDS, TELECOMM & ECON DEVELOPMENT
1977
HORIZON HOUSE INTL: DEDHAM, MA

MEXICO TELECOMM RURAL SOC
ECON TELEPHONE TELEX TV
SAT M-WAVE

929 VARIS, TAPIO
ASPECTS OF THE IMPACT OF TRANSNATIONAL COOPERATION ON COMMUNICATION
1981, SPRING
INTERNAT SOC SCIENCE J
28 N2 P809-830

INTERNAT ECON TECH SOC
CULTURE

930 VASUDIVAN, C.P.
MULTISTAGES AND SINGLE CHANNEL VHF FOR PUBLIC CALL OFFICES (PCSS) AND TELEPHONE SUBSCRIBERS IN OUTLYING AREAS ....
1979, APRIL
WORLD BANK: WASHINGTON, D.C.

TELEPHONE RURAL COST

931 VELASQUEZ, A. & ANGEL, D.
PLANNING THE OPTIMUM UTILIZATION OF PREASSIGNED AND DEMAND SERVICES IN DOMESTIC SATELLITE TELECOMMUNICATION SYSTEMS
TECH REP NO 5

TELECOMM SAT PLAN

932 VELASQUEZ, A. AND CASEY-STAHMER, A.
A PILOT PROJECT FOR RURAL COMMUNICATIONS IN PERU
1979
PAPER/ ACADEMY FOR EDUCATIONAL DEVELOPMENT, WASHINGTON, D.C.

PERU TELECOMM SAT RURAL
PLAN

933 VENUGOPAL, D. & NARAYANAN, K.
RURAL SATELLITE TELECOMM IN INDIA: APPROACH FOR SYSTEM OPTIMIZATION
1984
CANADIAN DOMESTIC & INTERNAT SAT COMM CONF 1ST OTTAWA, CANADA

INDIA SAT TELECOMM RURAL
TECHNICAL ASIA

934 VEST C.R.
EDUCATIONAL SATELLITES - SYSTEMS MODELS
1976
J OF TECHNICAL WRITING AND COMMUNICATIONS
6 N2 P145-62

EDUC SAT

935 VIERA-GALLO, JOSE ANTONIO
DOCUMENTATION AND THE DEMOCRATIZATION OF INFORMATION
1982
INTERNAT SOC SCIENCE J
34 N4
ERIC # EJ277254

SOC POLIT RES TECH
INFO TELECOMM DATA

936 VOLAN, CLAY
A HANDBOOK FOR THE STUDY OF SATELLITE COMMUNICATIONS IN A DEVELOPING WORLD
1981 OCT-NOV
EAST-WEST COMMUNICATION INSTITUTE: HONOLULU, HI

SAT RES

937 VON RABENAU, B., & STAHL, K.
DYNAMIC ASPECTS OF PUBLIC GOODS: A FURTHER ANALYSIS OF THE TELEPHONE SYSTEM
1974
BELL JOURNAL OF ECONOMICS
5 P651-669

EVAL ECON TELEPHONE

938 VORONOV, B. A.
THE CONSUMPTION OF COMMUNICATIONS OUTPUT
1969
VESTNIK SVIAZI (IN RUSSIAN)
8

RES ECON USSR

939 WA-WA-TA NATIVE COMMUNICATIONS SOCIETY
INTERVENTION BEFORE THE CANADIAN RADIO-TV AND
TELECOMMUNICATIONS COMMISSION
1978

CANADA TELECOMM POLICY RADIO
TV

940 WAD, ATUL
MICROELECTRONICS: IMPLICATIONS AND STRATEGIES FOR THE
THIRD WORLD
1982, OCT
THIRD WORLD QUARTERLY
4 N4 P677-697

COMPUTER INFO TECH POLICY
PLAN

941 WAG, G. AND DISSANAYAKE, W. (EDS.)
CONTINUITY AND CHANGE IN COMMUNICATION SYSTEMS: AN ASIAN
PERSPECTIVE
1985
ABLEX PUBLISHING: NORWOOD, NEW JERSEY

ASIA CHINA IRAN TAIWAN
MEDIA TV CULT

942 WALLENSTEIN, G.D.
GLOBAL SCOPE AT NATIONAL LEVEL - ITU'S CONTRIBUTIONS TO
TELECOMMUNICATION DEVELOPMENT
1976
IEEE TRANSACTIONS ON COMMUNICATIONS
24 N7 P700-9

INTERNAT TECH TELECOMM ITU
TECHNICAL

943 WALLENSTEIN, G.D.
HANDBOOKS OF CONSULTATIVE COMMITTEES -- BRIDGES BETWEEN
INTERNAT STANDARDIZATION AND NATIONAL TELECOMM DEVELOPMENT
1976
TELECOMMUNICATION JOURNAL
43 N1 633-7

TECH INTERNAT RURAL

944 WALP, R. M.
TELECOMM IN ALASKA. PAPERS IN SUPPORT OF THE ALASKA CASE STUDY PRESENTATION TO THE 1982 PAC TELECOMM CONFER
1982, JAN
PACIFIC TELECOMMUNICATION COUNCIL: HONOLULU

N AMER US ALASKA TELECOMM
SAT RURAL ECON SOC
TECHNICAL

945 WALP, R. M.
THE EVOLUTION, DESIGN AND USE OF A TELECOMMUNICATIONS SYSTEM IN A LIGHTLY POPULATED REGION
1977

ALASKA TELECOMM RURAL PLAN
TECH

946 WALSHAM, G.
TELEPHONE SYSTEMS IN THE THIRD WORLD; A RESEARCH PROJECT
1983, MARCH
THE J OF OPERATIONAL RES SOCIETY. OXFORD
34 N3 P225

TELEPHONE RES

947 WALTON, P.
BROADCASTING TECHNOLOGICAL AIDS TO THE THIRD WORLD
1984, JAN 19
COMPUTING (GREAT BRITAIN)

TELECOMM UK BUSI

948 WARDROP, D. H.
TELEPHONE ASSISTANCE PLAN
1978
CANADIAN DEPT OF COMMUNICATIONS: OTTAWA

CANADA TELEPHONE PLAN

949 WCISLY, K.
A NEW GENERATION OF MANAGEMENT SYSTEMS
1977
ZYCIE WARSZAWY (IN POLISH)
10

POLAND MGT TELECOMM

950 WCISLY, K.
DO YOU HAVE A TELEPHONE?
1977
ZYCIE GOSPODARCZE (IN POLISH)
40

POLAND TELEPHONE

951 WCISLY, K.
TELECOMMS-DEVELOPMENT FACTOR
1978
PRZAGLAD TECHNICZNY (IN POLISH)
10

POLAND TELECOMM

952 WCISLY, K.
THE ROLE OF TELECOMMUNICATIONS IN ECONOMIC GROWTH
1976
WYBRANE AKTUALNE PROBLEMY TELEKOMMUNIKACJU (IN POLISH)
INSTYTUT ZACZNOSCI: WARSAW, POLAND

POLAND TELECOMM ECON

953 WCISLY, K.
THE ROLE OF TELECOMMUNICATIONS IN SOCIO-ECONOMIC DEVELOPMENT:
A REVIEW OF EMPIRICAL STUDIES IN POLAND AND THE USSR
1978
MIMEO

USSR POLAND TELECOMM RES
STAT

954 WEBSTER, R. LYLE, (ED)
INTEGRATED COMMUNICATION: BRINGING PEOPLE AND RURAL DEVELOPMENT TOGETHER
1975
EAST-WEST COMMUNICATION INSTITUTE: HONOLULU, HAWAII

RURAL

955 WEDEMEYER, DANIEL J.
DEVELOPING COMMUNICATIONS SERVICES FOR HUMAN DEVELOPMENT
1980
MEDIA ASIA
7 N1 P41-66

TELECOMM POLICY

956 WEDEMEYER, DANIEL J. (ED)
PROCEEDINGS OF THE PACIFIC TELECOMMUNICATIONS CONFERENCE
1982, JAN
PACIFIC TELECOMMUNICATIONS COUNCIL: HONOLULU, HI

PAC	TELECOMM	L AMER	N AMER	ASIA
ALASKA	SAT	POLICY	INFO	FIBER
COMPUTER	TECH	AUSTRALIA	SOC	ECON

957 WEDEMEYER, DANIEL J. (ED)
PROCEEDINGS OF THE PACIFIC TELECOMMUNICATIONS CONFERENCE
1983, JAN
PACIFIC TELECOMMUNICATIONS COUNCIL: HONOLULU, HI

PAC	TELECOMM	L AMER	N AMER	ASIA
ALASKA	SAT	POLICY	INFO	FIBER
COMPUTER	TECH	AUSTRALIA	SOC	ECON

958 WEDEMEYER, DANIEL J. (ED)
PROCEEDINGS OF THE PACIFIC TELECOMMUNICATIONS CONFERENCE
1984, JAN
PACIFIC TELECOMMUNICATIONS COUNCIL: HONOLULU, HI

PAC	TELECOMM	L AMER	N AMER	ASIA
ALASKA	SAT	POLICY	INFO	FIBER
COMPUTER	TECH	AUSTRALIA	SOC	ECON

959 WEDEMEYER, DANIEL J. (ED)
PROCEEDINGS OF THE PACIFIC TELECOMMUNICATIONS CONFERENCE
1985, JAN
PACIFIC TELECOMMUNICATIONS COUNCIL: HONOLULU, HI

PAC	TELECOMM	L AMER	N AMER	ASIA
ALASKA	SAT	POLICY	INFO	FIBER
COMPUTER	TECH	AUSTRALIA	SOC	ECON

960 WEDEMEYER, DANIEL J. (ED)
PROCEEDINGS OF THE PACIFIC TELECOMMUNICATIONS CONFERENCE
1986, JAN
PACIFIC TELECOMMUNICATIONS COUNCIL: HONOLULU, HI

PAC	TELECOMM	L AMER	N AMER	ASIA
ALASKA	SAT	POLICY	INFO	FIBER
COMPUTER	TECH	AUSTRALIA	SOC	ECON

961 WELDOM, K.L., SCHULMAN, AND LESSER, B.
ECONOMIC BENEFITS OF IMPROVED TELEPHONE SERVICE TO RURAL AREAS:
PHASE I REPORT
1977
UNPUBLISHED REPORT
DALHOUSIE UNIVERSITY: HALIFAX, NOVA SCOTIA

ECON TELEPHONE RURAL

962 WELLENIUS, BJORN
ECONOMIC ANALYSIS OF TELECOMMUNICATIONS INVESTMENT IN DEVELOPING
COUNTRIES (WORLD TELECOMMUNICATIONS FORUM. PART I.)
1983
SYMPOSIUM ON POLICY, FIN AND ECO ASPECTS OF TELECOMM (ITU)
INTERNATIONAL TELECOMMUNICATION UNION: GENEVA

TELECOMM FIN ECON

963 WELLENIUS, BJORN
ESTIMATING TELEPHONE NECESSITIES IN DEVELOPING NATIONS
1971
IEEE TRANSACTIONS ON COMMUNICATION TECH
19 P333-39

TELEPHONE POLICY SOC

964 WELLENIUS, BJORN
ON THE ROLE OF TELECOMMUNICATION IN DEVELOPMENT
1984, MARCH
TELECOMM POLICY
8 N1 59-66

TELECOMM EVAL THEORY

965 WELLENIUS, BJORN
RURAL TELECOMMUNICATION IN THE DEVELOPING COUNTRIES: INTRODUCTION
IN POLISHUK & O'BRYANT, EDS, TELECOMM & ECON DEVELOPMENT
1977
HORIZON HOUSE INTL: DEDHAM, MA

TELECOMM ECON RURAL

966 WELLENIUS, BJORN
SOME RECURRENT PROBLEMS OF TELECOMMUNICATION IN DEVELOPING COUNTRIES
1976
IEEE TRANSACTIONS ON COMMUNICATIONS
24 N7 P723-83-

TELECOMM ITU ECON FIN

967 WELLENIUS, BJORN
TELECOMMUNICATION IN DEVELOPING COUNTRIES
1984, SEPT
FINANCE & DEVELOPMENT
21 N3 P33-36

TELECOMM

968 WELLENIUS, BJORN
TELECOMMUNICATIONS IN DEVELOPING COUNTRIES
1977, SEPT
TELECOMM POLICY
I P289-297

TELECOMM

969 WELLENIUS, BJORN
THE ROLE OF TELECOMMUNICATION SERVICES IN DEVELOPING COUNTRIES
1978
CANADIAN DEPT OF COMMUNICATIONS: OTTAWA

TELECOMM EVAL

970 WELLENIUS, BJORN
THE ROLE OF TELECOMMUNICATIONS IN THE DEVELOPMENT OF NATIONS
1984, MARCH
TELECOMMUNICATION POLICY
P59-66

TELECOMM ECON

971 WELLENIUS, BJORN, CASTILLO, L. & MELNICK, E.
METODOLOGIA PARA EL ESTUDIO DE LAS TELECOMMUNICACIONES RURALES:
CUATRO CONTRIBUCIONES
1971
RES REP NO 28

CHILE RURAL RES TELECOMM

972 WELLENIUS, BJORN, WEBBER, M., CLEEVELY, D. D.
ON THE ROLE OF TELECOMMUNICATION IN DEVELOPMENT/
REJOINERS
1984, MARCH
TELECOMM POLICY

TELECOMM

973 WELLENIUS, BJORN
ON THE ROLE OF TELECOMMUNICATIONS IN THE DEVELOPMENT
OF NATIONS
1972
IEEE TRANSACTIONS ON COMMUNICATION TECH
20 P3-8

TELECOMM ECON EVAL

974 WELLS, STUART
INSTRUCTIONAL TECHNOLOGY IN DEVELOPING COUNTRIES: DECISION-MAKING PROCESS IN EDUCATION
1976
PRAEGER PUB INC.: NEW YORK

TECH EDUC

975 WELLS, STUART AND KLEES, STEVEN
EDUCATION DECISIONS AND COST ANALYSIS: THE RADIO MATHEMATICS PROJECT IN NICARAGUA
1977, FEBRUARY
PAPER/U.S. AGENCY FOR INTERNATIONAL DEVELOPMENT

NICARAGUA RADIO EDUC COST

976 WEST, ANITA S.
DESIGN ISSUES FOR DEMONSTRATION OF RURAL SERVICE DELIVERY PROJECTS INCORPORATING TELECOMMUNICATION TECHNOLOGIES
1978, SEPT
PB 290 221/1 & REPT NSF/RA-780312

RURAL TELECOMM MODEL

977 WESTPHAL, LARRY
RESEARCH ON APPROPRIATE TECHNOLOGY
1978
INDUSTRY AND DEVELOPMENT
2 P28-46
NY

INTERNAT TECH RES POLICY

978 WHITE, C.E.
TELECOMMUNICATION NEEDS OF DEVELOPING COUNTRIES
1976, SEPTEMBER
TELECOMMUNICATION POLICY
P19-23

TELECOMM

979 WHITE, KATHLEEN M. & JACOBSON, C. RANDALL
TRADE ISSUES IN TELECOMMUNICATION & INFORMATION, VIII. TRADE
IN TELECOMMUNICATION AND INFO PRODUCTS WITH DEVELOPING COUNTRIES
1981, APRIL
PB81-216301

TELECOMM BUSI INFO ECON

980 WHITE, P.B. AND KELABORA, L.
COMMUNICATION SATELLITES AND EDUCATION IN INDONESIA: WHAT IS AN
APPROPRIATE STRATEGY?
1978, JUNE
EDUCATIONAL BROADCASTING INTERNATIONAL
11 N2 P95-99

INDONESIA SAT EDUC RADIO
TV

981 WHITE, PETER B.
AUSTRALIA'S DOMESTIC COMMUNICATION SATELLITE AND EDUCATION:
HAS EDUCATION MISSED THE BOAT?
ED245673
ERIC DOC REPRO SERV

AUSTRALIA SAT EDUC

982 WHITE, ROBERT AND MCDONNELL, JAMES
PRIORITIES FOR NATIONAL COMMUNICATION POLICY IN THE THIRD
WORLD
1983
THE INFO SOCIETY J
2 N1 P5-33

TELECOMM MASS MEDIA INTERNAT POLICY

983 WHYTE, J. S.
TELECOMMUNICATIONS IN THE SERVICE OF MAN
1971
ICC CONF PAPER

WORLD TELECOMM SOC

984 WHYTH, S.E. & STEINES, L.W.
TELECOMMUNICATION DEVELOPMENTS IN LATIN AMERICA
IN THE CURRENT DECADE
1981, JUNE
COMMUNICATION INTERNATIONAL (GB)
8 N6

L AMER TELECOMM TECHNICAL

985 WIGAND, ROLF T.
DIRECT SATELLITE BROADCASTING: SELECTED SOCIAL IMPLICATIONS. IN
COMMUNICATIONS YEARBOOK 6, BURGOON, M. (ED.)
1982
P250-288
SAGE PUBLICATIONS: BEVERLY HILLS, CA

TECH SAT SOC TV
DATA MGT

986 WIGAND, ROLF T.
THE DIRECT SATELLITE CONNECTION: DEFINITIONS AND PROSPECTS
1980, SPRING
JOURNAL OF COMMUNICATION
30 N2 P140-156

SAT

987 WIGAND, ROLF T. ET AL
TRANSBORDER DATA FLOW, INFORMATICS AND NATIONAL POLICIES:
A COMPARISON AMONG 22 NATIONS
ED235530
ERIC DOC REPRO SERV

WORLD INFO TELECOMM POLICY

988 WILDE, GLENN R.
INTERNATIONAL COMMUNITY LEARNING AND INFORMATION CENTERS:
CONCEPT, PROCESS AND PROPOSAL
1984
ASSN PAPER

EDUC INFO

989 WILDER, W.D.
COMMUNICATION, SOCIAL STRUCTURE AND DEVELOPMENT IN RURAL MALAYSIA: A STUDY OF KAMPUNG KUALA BERA
LONDON SCHOOL OF ECONOMICS MONOGRAPHS ON SOCIAL ANTHROPOLOGY
N56
THE ATHALON PRESS: LONDON

MALAYSIA ASIA MEDIA CULT
RURAL

990 WILKENS, H., AND PLENGE, G.
TELECONFERENCE DESIGN: A TECHNICAL APPROACH TO SATISFACTION
1981, SEPTEMBER
TELECOMMUNICATIONS POLICY
5 N3 P216-227

TELECONF TECHNICAL

991 WILLIAMS, HAROLD
ELECTRONIC MAIL: ADVANTAGES, SERVICES, AND CONTROLS
1982, NOVEMBER
TCS TELECOMMUNICATIONS
P63-66

COMPUTER

992 WILLIAMSON, JOHN
ASIA TELECOMMUNICATION '85: RESPONSE TO A GROWING NEED
1986, MARCH 25
TELEPHONY
P72-74

ASIA TELECOMM

993 WILLIAMSON, JOHN
TELECOMMUNICATION EXPANSION IN CHINA'S ECONOMIC ZONES
1985, APRIL 22
TELEPHONY
208 N16 P61-5

CHINA ECON TELECOMM PLAN

994 WILLIAMSON,JOHN
MECOM'85: MID-EAST MARKET MATURES
MIDDLE EAST ELECTRONIC COMMUNICATIONS EXPOSITION/BAHRAIN
1985, MARCH 25
TELEPHONY
P68-71

M EAST FIN ECON

995 WINHAM, G. R.
POLITICAL DEVELOPMENT AND LERNER'S THEORY:
FURTHER TEST OF A CAUSAL MODEL
1970
AMER POLIT SCIENCE REVIEW
64 P810-18

POLIT MODEL THEORY

996 WITHAM, A., HAWKER, P.
THE DEVELOPMENT OF SATELLITE BROADCASTING
N 1980
INT. BROADCAST ENG (GB)
1 N174 P44-9

HIST TECH SAT BROADCAST
TV UK

997 WOLF, NANCY E. AND HOBBS, DAVID B.
TRADE ISSUES IN TELECOMM AND INFO: THE ROLE OF US GOVERNMENT
IN FINANCING OF EXPORTS IN MAJOR TELECOMM AND INFO PRODUCTS
1981 JUNE
PB81-240616

US FIN POLIT POLICY
TELECOMM INFO BUSI

998 WOLTER, W.
IMPROVED TELECOMMUNICATIONS-KEY TO WORLD ECONOMIC GROWTH
1977
TELEPHONY
192 (20) P86-88

TELECOMM WORLD ECON

999 WOLTER, W.G.G.
TELECOMMUNICATION: LIFE BLOOD OF ECONOMY, SOCIETY
1980, DEC
TELEPH. ENG. AND MGT
84 N24

TELECOMM ECON SOC

1000 WOOD, F.B., COATES,V.T., CHARTRAND,R.L., ERICSON, R.F.
VIDEOCONFERENCE VIA SATELLITE - OPENING GOVERNMENT TO PEOPLE
1978
FUTURIST
12 N5 P321

SAT TELECONF POLIT

1001 WOODFORD, J.B.
COMMUNICATION SATELLITE TECHNOLOGY: STATE OF THE ART & DEVELOPMENT OPPORTUNITIES
1978 JULY
NTIS N80 25358/6

SAT TECH EVAL

1002 WORLD BANK
TELECOMMUNICATIONS SECTOR WORKING PAPERS
1971 TO PRESENT
WORLD BANK: WASHINGTON, D.C.

WORLD BANK TELECOMM PLAN ECON
TECH POLICY

1003 WORLD BANK
WORLD DEVELOPMENT REPORT
ANNUAL
WORLD BANK: WASHINGTON, D.C.

INTERNAT WORLD BANK STAT ECON

1004 WORLD BANK
WORLD TABLES
1983
WORLD BANK: WASHINGTON, D.C.

INTERNAT WORLD BANK STAT ECON

1005 WU, KAIZHEN W. AND MILLER, MICHAEL J.
POINT-TO-MULTIPOINT DIGITAL SUBSCRIBER RADIO SYSTEMS FOR RURAL AREAS IN CHINA
1985
PAPER/ PACIFIC TELECOMMUNICATIONS FOR PACIFIC DEVELOPMENT
P269-274

CHINA RADIO DIGITAL TELEPHONE
RURAL ECON REMOTE AUSTRALIA
FRANCE JAPAN

1006 WUNDER, BERNARD J.
LONG-RANGE GOALS IN INTERNATIONAL TELECOMM AND INFORMATION: AN OUTLINE FOR UNITED STATES POLICY
1983, FEBRUARY
REPORT SUBMITTED TO THE CONGRESS OF THE UNITED STATES

INTERNAT US TELECOMM INFO
POLICY ITU TECH FIN
RES

1007 WURTZEL, A. H. & TURNER, C.
LATENT FUNCTIONS OF THE TELEPHONE: WHAT MISSING THE EXTENSION MEANS
1977
IN THE SOC IMPACT OF THE TELEPHONE, I. POOL (ED)

TELEPHONE HIST

1008 WYTHE, S.E. AND STEINES, L.
TELECOMMUNICATIONS DEVELOPMENT IN LATIN AMERICA AND IN CURRENT DECADE
1981, MAY
COMMUNICATIONS INTERNATIONAL
8 N5

L AMER TELECOMM

1009 YOUNG, ELIZABETH L., AND HURD, JANE
PACIFIC BASIN COMMUNICATIONS STUDY VOL. 1 AND 2
1981
PUBLIC SERVICE SATELLITE CONSORTIUM: WASHINGTON, D.C.

PAC SAT

1010 YOUNG, P., OWUSU, R., MCMORRIN, G.
AFRICAN TELECOMMUNICATIONS
1982, JULY
AFRICA
N 131 P73

AF TELECOMM

1011 YOUNG, PETER
EUROPE LEADS AFRICA FOLLOWS
1981, JANUARY
AFRICAN BUSINESS
P41-42

EUR AF TELECOMM

1012 YOUNG, PETER
TELECOMMUNICATIONS SURVEY (AFRICAN TELECOMM SYSTEMS)
1980, FEB
AFRICAN BUSINESS
P33-5

AF TELECOMM INFO

1013 YUROW, JANE H.(ED)
ISSUES IN INTERNATIONAL TELECOMMUNICATION POLICY:
A SOURCEBOOK
1983
CENTER FOR TELECOMM STUDIES, GEORGE WAHSINGTON UNIV.: WASHINGTON,DC

INTERNAT TELECOMM POLICY

1014 ZIGERELL, JAMES
DISTANCE EDUCATION: AN INFORMATION AGE APPROACH TO ADULT EDUCATION
ED246311
ERIC DOC REPRO SERV

EUR TELECOMM SAT EDUC
TV

1015 ZWIMPFER, LAURENCE E.
EMERGING INFORMATION SERVICES IN NEW ZEALAND. IN TELECOMMUNICATIONS FOR PACIFIC DEVELOPMENT, WEDEMEYER, D.J. (ED.)
1985
P 14-21

NEW ZEALAND TELECOMM DIGITAL V-TEXT
T-TEXT

1016 A HANDBOOK OF INTELSAT: BUSINESS PLANNING AND SERVICE DEVELOPMENT (REVISION)
1985, FEB 28
INTELSAT

INTELSAT BUSI PLAN

1017 A STUDY ON THE PACIFIC REGIONAL SATELLITE COMMUNICATIONS SYSTEM
1983
PAPER/RESEARCH INSTITUTE OF TELECOMMUNICATIONS AND ECONOMICS (RITE)
PACIFIC TELECOMMUNICATION CONFERENCE '83: HONOLULU, HAWAII

SAT PAC

1018 AFRICA'S TELECOMMUNICATION CHALLENGE: A SPECIAL SURVEY
1983, JAN
AFRICAN BUSINESS
P45

AF TELECOMM ECON

1019 AFRICAN TELECOMMUNICATION
1980, FEB
AFRICA (AFN J, LTD)
P63773,

AF TELECOMM ECON

1020 ANALYSIS OF THE EXPORT DEVELOPMENT POTENTIAL OF THE US TELECOMM EQUIPMENT INDUSTRY
1982 AUG
ITA-82-11-602

EVAL US TELECOMM TECH
BUSI

1021 APPROPRIATE MODERN TELECOMMUNICATIONS TECHNOLOGY FOR INTEGRATED RURAL DEVELOPMENT IN AFRICA
1981, NOVEMBER
INTERNATIONAL TELECOMMUNICATION UNION: GENEVA, SWITZERLAND

AF TELECOMM TECH RURAL
ITU

1022 ARGENTINA BUYS HARRIS EARTH STATIONS IN NATIONAL SATELLITE NETWORK
1981, JULY
COMMUNICATIONS INTERNATIONAL
P10

ARGENTINA SAT L AMER

1023 BARRIERS TO TELECOMMUNICATION & INFORMATION TRADE
1984, AP-MAY
TRANSNATL DATA REP (NETHERLANDS)
7 N3 P144-9

TELECOMM INFO BUSI INTERNAT

1024 BUILDING A PAN AFRICAN COMMUNICATION NETWORK
1972 APR
AFRICA REPORT
17 N4

AF PANAFTEL TECH TELECOMM

1025 CHANGES IN THE US TELECOMMUNICATION INDUSTRY & THE IMPACT ON US TELCOMM TRADE; REPORT TO COMMITTEE ON FINANCE, US SENATE
1984
GPO ITEM 980-EOFICHE

US TELECOMM LEGAL POLICY
BUSI EVAL

1026 COMMUNICATION IN THE SERVICE OF MAN
1983 MARCH
UNESCO COURIER (COMMUNICATIONS)
36 N3
ERIC # EJ279817

COMPUTER TELECOMM CULT UNESCO
MEDIA EDUC

1027 COMMUNICATION TECHNOLOGY IN EDUCATION AND TRAINING
1982, MAR 15-17
PROCEEDING/ CONF ON COMM TECHNOLOGY IN EDUCATION AND TRAINING
INFORMATION DYNAMICS INC.: MARYLAND

TECH EDUC TRAIN

1028 COMMUNICATIONS IN THE PACIFIC BASIN
1980, APRIL
PAPER/ PUBLIC SERVICE SATELLITE CONSORTIUM
COMSAT GENERAL CORPORATION: WASHINGTON, D.C.

PAC SAT VIDEO TELEPHONE

1029 DEVELOPMENT & EVOLUTION OF THE TELEPHONE SERVICE
1983, DEC
TELECOMM J
50 N12 679-82

TELEPHONE HIST

1030 DEVELOPMENT COMMUNICATION REPORT
1983 MARCH
N 41
ERIC #ED231330

HONDURAS SOUTH PAC NEPAL US
RADIO EDUC HEALTH COMPUTER
TELECOMM

1031 DEVELOPMENT COMMUNICATION REPORT
1985, WINTER
ERIC DOC REPRO
N 48 DOC REPRO
ED257413

TELECOMM SAT EDUC HEALTH
AGRIC RURAL MEDIA

1032 DEVELOPMENT COMMUNICATIONS IN THE THIRD WORLD
PROCEEDINGS OF A MIDWEST REGIONAL SYMPOSIUM
1983, APRIL
UNIVERSITY OF ILLINOIS AT URBANA-CHAMPAIGN

TELECOMM

1033 DEVELOPMENT OF TELECOMMUNICATIONS IN SOUTH ASIA AND CHINA
1981
TELECOMMUNICATION JOURNAL
48 N6 P292-313

SOUTH ASIA CHINA TELECOMM

1034 DISCUSSION SEMINAR ON TWO-WAY FOR RURAL HEALTH SERVICES IN DEVELOPING COUNTRIES
1980, JUNE 4-5
BOARD ON SCIENCE AND TECHNOLOGY FOR INTERN. DEVELOPMENT
WASHINGTON, D.C.

HEALTH TECH RURAL TRAIN

1035 DOMESTIC SPACE COMMUNICATION: THE INDONESIAN EXPERIENCE
1982, JAN 19-27
PAPER/ECONOMIC AND SOCIAL COMMISSION FOR ASIA AND THE PACIFIC/ITU

INDONESIA SAT M-WAVE SOC
ECON POLIT

1036 ECONOMIC IMPROVEMENT PUTS FRESH GLOW ON GROWTH OF TELECOMM MARKET
1984, JULY
OFFICE EQUIPMENT AND PRODUCTS
13 N69 P58-9

ECON TELECOMM BUSI EVAL

1037 ECONOMICS IN TELECOMMUNICATION, COMPUTERS & ELECTRONICS; EDUCATIONAL COMPONENT: HEARING BEFORE SUBCOMM ON TELECOMM, CONSUMER PROJECT
1984
SERIAL # 98-91 GPO ITEM 1019-A, 1019-B

ECON TELECOMM COMPUTER TECH
POLIT POLICY

1038 ECOWAS AND TELECOMMUNICATION
1981
WEST AFRICA
1:1

FIN AF ECOWAS TELECOMM

1039 ECOWAS/PANAFTEL:FEASIBILITY STUDY
1980
WEST AFRICA
2:25

PANAFTEL ECOWAS AF SAT

1040 EDUCATION IN LATIN AMERICA
1983 SPRING
J OF CONTINUING HIGHER EDUCATION
V31 N2 P22-5

EDUC L AMER TELECOMM RURAL

1041 EDUCATIONAL TELECOMMUNICATION FOR ALASKA:
VOLUME III: ALASKA KNOWLEDGE BASED SYSTEM
ALASKA STATE DEPARTMENT OF EDUCATION: JUNEAU, ALASKA

TELECOMM EDUC INFO RURAL
COST DATA

1042 EDUTEL, COMMUNICATION AND DEVELOPMENT
COMPUTERIZED TELECOMMUNICATION PLANNING TOOLS
1978
ACADEMY FOR EDUCATION DEVELOPMENT: WASHINGTON, D.C.

TELECOMM PLAN TECHNICAL RES

1043 ESTABLISHING A TELECOMMUNICATION TRAINING CENTER IN A DEVELOPING COUNTRY
1981, JUNE
TELECOMMUNICATION JOURNAL
48,N6, P304-313

TELECOMM TRAIN

1044 FAX NETWORK GIVES DOCTORS IMMEDIATE ACCESS TO MEDICAL SCHOOL RESOURCES
1983, JANUARY
COMMUNICATIONS NEWS
P64-65

HEALTH TECH

1045 FEASIBILITY AND PLANNING OF INSTRUCTIONAL SATELLITE SYSTEMS
1982, AUG 9-12
REPORT/U.N. CONF ON EXPLORATION AND PEACEFUL USES OF OUTER SPACE
VIENNA, AUSTRIA

SAT EDUC PLAN TECHNICAL
SOC ECON ORBIT MGT
EVAL INDIA

1046 FINANCING TELECOMMUNICATIONS FOR RURAL DEVELOPMENT
1984, SEPT
TELECOMM POLICY
8 N 3 P181

FIN RURAL TELECOMM INVEST

1047 FOR TELECOMMUNICATIONS DEVELOPMENT: US 2.39 BILLION AT ITS DISPOSAL
1978, SEPT
MALAYSIAN BUSI, KUALA LUMPUR
P21

TELECOMM US INVEST

1048 IMPACT OF MODERN COMMUNICATION TECHNOLOGY-II: INDONESIA
1981
DOCUMENTS OF THE NEW COMMUNICATION ORDER NO. 2
UNESCO: PARIS, FRANCE

INDONESIA MEDIA SAT ECON
SOC CULT INFO RURAL

1049 IMPACT OF MODERN COMMUNICATION TECHNOLOGY
1981
DOCUMENTS OF THE NEW COMMUNICATION ORDER
UNESCO: PARIS, FRANCE

SOC INTERNAT ECON POLICY CULT MEDIA SAT

1050 INDONESIAN TELECOMMUNICATIONS PUBLIC CORPORATIONS
1981
REPORT/REGIONAL TELECOMM TRAINING MANAGEMENT MEETING/ JAKARTA
P1-16

INDONESIA TELECOMM TRAIN

1051 INFORMATION TECHNOLOGY FOR AGRICULTURAL AMERICA.
PREPARED FOR SUBCOMM ON DEPARTMENT OPERATIONS, RESEARCH & FOR AG
97TH CONGRESS, 2ND SESSION
ED239585
ERIC DOC REPRO SERV

AGRIC US TELECOMM COMPUTER

1052 INTELCOM '80 CONFERENCE PROCEEDINGS: COMPETITION IN THE INFORMATION ECONOMY
1980
ARTECH HOUSE: DEDHAM, MA

TELECOMM SAT TECHNICAL PLAN TV CABLE FIBER DIGITAL RADIO FIN

1053 INTERAGENCY REPORT ON RURAL COMMUNICATIONS
OFFICE OF TELECOMMUNICATIONS POLICY
1977, DEC 15
EXECUTIVE OFFICE OF THE PRESIDENT

TELECOMM POLICY RURAL

1054 INTERNAT TELECOMM & INFO POLICY: HEARINGS BEFORE SUBCOMMITTEE
OF COMMITTEE ON GOV OPERATIONS, HOUSE OF REP., 97TH CONGRESS
1982
GPO ITEM 1016-A

INTERNAT TELECOMM INFO POLICY
US

1055 INTERNATIONAL SATELLITE ISSUES HEARINGS, SUBCOMM ON TELECOMM
CONSUMER PROTECTION HSE OF REP 98TH CONGRESS 2ND SESSION
1985
GPO ITEM 1019-A, 1019-B (MICROFICHE)
N98-177
GPO WASH D.C.

INTERNAT SAT POLICY TELECOMM
US LEGAL

1056 INTERNATIONAL TELECOMMUNICATION POLICIES: HEARING BEFORE SUBCOMM OF
COMM ON COMMERCE, SCIENCE, & TRANSPORTATION US SENATE 95TH CONGRESS
1978
GPO ITEM 1041

INTERNAT POLICY TELECOMM US

1057 INTERNATIONAL TRADE ISSUES IN TELECOMM & RELATED INDUSTRIES; HEARING
SUBCOMM ON TELECOMM, CONSUMER PROTECTION & FINANCE 97TH CON
1983
GPO ITEM 1019-A, 1019-B (MICROFICHE)

INTERNAT TELECOMM BUSI US
LEGAL

1058 ITU-OECD: TELECOMMUNICATION & RURAL DEVELOPMENT
1984
AGORA INF CHANGING WORLD (ITALY)
N7 P34-6

ITU TELECOMM RURAL

1059 JOINT CONGRESSIONAL HEARING ON THE CHANGING INFO NEEDS OF RURAL AMERICA: THE ROLE OF LIBRARIES AND INFORMATION TECHNOLOGY
ED247958
ERIC DOC REPRO
US

COMPUTER INFO TELECOMM RURAL

1060 LONG-RANGE GOALS IN INTERNATIONAL TELECOMMUNICATION AND INFORMATION: AN OUTLINE FOR UNITED STATES POLICY
ED244604
ERIC DOC REPRO SERV

US TELECOMM INFO INTERNAT
POLICY PLAN

1061 MAINTAINING THE NETWORK: RURAL TELEPHONE SERVICE & FCC ACTIONS: 43RD REPORT
1982
GPO ITEM 1008-C 1008-D (MICROFICHE)

RURAL TELEPHONE LEGAL POLICY
US

1062 PACSAT: LOW-COST SATELLITE COMMUNICATION FOR DEVELOPMENT
1986, JANUARY
REPORT/ INTERN. PROJECT OF VOLUNTEERS IN TECHNICAL ASSISTANCE

SAT COST PAC

1063 PACSAT: SATELLITE COMMUNICATION
1984, APRIL
VITA NEWS
P4-9

SAT PAC

1064 PADIS (PAN AFRICAN DOCUMENTATION AND INFORMATION SYSTEM)
UNITED NATIONS ECONOMIC COMMISSION FOR AFRICA
1981, JANUARY
ADDIS ABABA, ETHIOPIA

AF INTERNAT INFO

1065 PANAFTEL (PAN AFRICAN TELECOMMUNICATION) NETWORK
1983
INTERNATIONAL TELECOMMUNICATION UNION: GENEVA, SWITZERLAND

AF TELECOMM ITU

1066 PHILIPPINE TELECOMMUNICATION POLICY DEVELOPMENT. IN TELECOMM PACIFIC
RAHIM, S.A., AND WEDEMEYER, D.J., (EDS.)
1983
P 137-143
PACIFIC TELECOMMUNICATIONS COUNCIL: HONOLULU, HAWAII

PHILIPPINES TELECOMM

1067 PROBLEMS CAUSED BY INADEQUATE COMMUNICATION SYSTEMS ON THE
INTERNATIONAL NETWORK
1985
THIRD WORLD DEVELOPMENT
2

INTERNAT

1068 PROPOSAL FOR A RURAL SATELLITE PROJECT WITH THE EASTERN ISLANDS
UNIVERSITY ASSOCIATION, INDONESIA
1981, SEPT 25
PAPER/ ACADEMY FOR EDUCATIONAL DEVELOPMENT, WASHINGTON, D.C.

INDONESIA SAT RURAL

1069 PROPOSAL TO CONDUCT FEASIBILITY STUDY OF SOFTWARE REQUIREMENTS AND GROUND SEGMENT DESIGN FOR PAKISTAN SATELLITE SYSTEMS
1983, SEPTEMBER
ACADEMY FOR EDUCATIONAL DEVELOPMENT: WASHINGTON, D.C.

PAKISTAN SAT TELECOMM

1070 PTC'85 TOWARD A DIGITAL WORLD. TELECOMM FOR PACIFIC DEVELOPMENT
PROCEEDINGS OF ANNUAL CONFERENCE OF PACIFIC TELECOMM COUNCIL
1985, JAN
REP # ISBN-0-8248-1009-0
ED-253 218
ERIC DOC REPRO SERVICE

PAC RES TELECOMM

1071 RADIO FOR EDUCATION AND DEVELOPMENT: CASE STUDIES
1977
1 AND 2
WORLD BANK PUBLICATION: WASHINGTON, D.C.

RADIO EDUC WORLD BANK

1072 RAPPORT DE LA COMMISSION TRANSPORTS ET COMMUNICATIONS, PREPARATION DU VIIEME PLAN
PARIS: COMMISSARIAT GENERAL DU PLAN. DOCUMENTATION FRANCAISE

1073 REA FINANCING OF TELECOMMUNICATIONS FACILITIES
HEARING BEFORE SUBCOMMITTEE ON AGRICULTURE CREDIT
1979
LCCN GP 79023326
GPO 1032-B
GPO, WASH D.C.

TELECOMM FIN RURAL AGRIC

1074 REA LOANS & LOAN GUARANTEES FOR RURAL ELECTRIC & TELEPHONE
SERVICE
1983
GPO ITEM 116

RURAL TELEPHONE FIN US
LEGAL POLICY

1075 REGIONAL CO-OPERATION FOR EDUCATION AND DEVELOPMENT IN AFRICA USING
SPACE COMMUNICATION
1977 APR
COMBROAD
N35 18-23

EDUC ECON AF UNESCO
ITU SAT TELEPHONE TV

1076 REPORT OF THE UN ECON COMMISSION FOR AFRICA REGIONAL SEMINAR
ON REMOTE SENSING APPLICATIONS AND SAT COMM FOR EDUC AND DEV
1981, JUNE
UN CONF ON EXPLORATORY AND PEACEFUL USES OF SPACE
UNITED NATIONS: NY

AF SAT EDUC ECON
TECHNICAL

1077 REPORT ON THE FIRST REGIONAL TELECOMMUNICATION TRAINING MANAGEMENT
MEETING
1981, SEPT-OCT
REPORT/ INDONESIAN TELECOMM PUBLIC CORPORATION (PERUMTEL)/ JAKARTA

INDONESIA TELECOMM TRAIN

1078 RESEARCH & DEVELOPMENT IN ELECTRONICS & TELECOMMUNICATION
1983, DEC
TELECOMM J
50 N12 667-72

RES TELECOMM TECHNICAL TECH

1079 RESEARCH AND DEVELOPMENT IN TELECOMMUNICATION IN DEVELOPING
COUNTRIES
1983
TELECOMMUNICATION JOURNAL
50 N3 P140-146

TELECOMM RES

1080 RESEARCH AND DEVELOPMENT IN TELECOMMUNICATION IN DEVELOPING
COUNTRIES
1982, JAN 19-27
ITU REGIONAL CONF ON DEV AND MGT OF TELECOMM IN ASIA AND PACIFIC
ECONOMIC AND SOCIAL COMM FOR ASIA AND THE PACIFIC AND ITU: BANGKOK

ASIA PAC TELECOMM RES

1081 RURAL ELECTRIFICATION & TELEPHONE REVOLVING FUND SELF-
SUFFICIENCY ACT OF 1984
1984
GPO ITEM 1008-C, 1008-D (MICROFICHE)

RURAL TELEPHONE FIN COST
POLICY LEGAL

1082 RURAL TELECOMM: HEARING BEFORE SUBCOMMITTEE ON COMM OF COMMITTEE
ON COMMERCE, SCIENCE & TRANSP US SENATE 97TH CONGRESS
1981
GPO IIEM 1041-A

RURAL TELECOMM US LEGAL
POLICY

1083 SATELLITE COMMUNICATIONS- THE SOCIAL BENEFITS
1985
THIRD WORLD DEVELOPMENT
2 P72-77

SAT SOC

1084 SATELLITE FOR RURAL DEVELOPMENT
1979
J OF COMMUNICATION
29 N4

SAT RURAL

1085 SATELLITE OPTIONS FOR RURAL TELECOMMUNICATIONS IN DEVELOPING COUNTRIES
1981, JUNE
APPLICATIONS MANAGEMENT, U.S. AGENCY FOR INTERNATIONAL DEVELOPMENT

SAT	RURAL	TELECOMM	HEALTH
AGRIC	TECHNICAL	AID	POLICY
PLAN	COST		

1086 SATELLITES AND THE TELECOMMUNICATION NEEDS OF THE SOUTH PACIFIC COUNTRIES
1980, MAR/APR
PAPER/ INTERNATIONAL TELECOMMUNICATION UNION

SOUTH PAC	TELECOMM	SAT	RURAL
AUSTRALIA	TECHNICAL		

1087 SELECTED FUNDING SOURCES FOR TELECOMMUNICATION PROJECTS
WINTER 1983-84
URBAN LEAGUE REVIEW
8 N1
ERIC # EJ294307

BUSI	TELECOMM	FIN	POLITICAL

1088 SEMINAR ON 2-WAY FOR RURAL HEALTH SERVICES IN DEVELOPING COUNTRIES
BOARD ON SCIENCE AND TECH FOR INTERNATIONAL DEVELOPMENT
1980, JUNE 4-5

RURAL	HEALTH	EVAL	TRAIN

1089 SHARING TELECOMMUNICATIONS TECHNOLOGY WITH LDC'S
1985
THIRD WORLD DEVELOPMENT
2 P41-49

TELECOMM TECH

1090 SOUTH PACIFIC TTC INAUGURATED IN FIJI
1981
TELECOMMUNICATION JOURNAL
48 N11 P628

SOUTH PAC FIJI

1091 SOVIET STUDIES ON THE INDIRECT BENEFITS OF INVESTMENT IN
TELECOMMUNICATION
1978
MIMEO

SOVIET TELECOMM INVEST

1092 SPACE RADIOCOMMUNICATIONS SYSTEM FOR AID FOLLOWING
NATURAL DISASTERS
1975
ITU BOOKLET N18

SAT TELECOMM HEALTH

1093 SPLITTING THE HEAVENS (POSSIBLE USES OF SAT BY DEVELOPING COUNTRIES,
DISPUTES BTWN DEV. AND DEVELOPING COUNTRIES OVER CURRENT USES)
S 1982
SOUTH
P8-14

SAT TECH POLICY

1094 STATISTICAL REPORT TELEPHONE BORROWERS/US DEPT OF AGRICULTURE, RURAL ELECTRIFICATION ADMINISTRATION
1980
GPO ITEM 115 (MICROFICHE)

RURAL TELEPHONE BUSI US
STAT FIN

1095 TELECOMM & INFORMATION PRODUCTS & SERVICES IN INTERNATIONAL TRADE: HEARINGS BEFORE SUBCOMMITTEE ON TELECOMM, CONSUMER PROTECTION
1981
GPO ITEM 1019-A, 1019-B (MICROFICHE)

TELECOMM INFO INTERNAT BUSI
US LEGAL

1096 TELECOMM & ITS IMPACT ON THE NEW ENGLAND ECONOMY: HEARING BEFORE THE SUBCOMMITTEE ON TELECOMM, CONSUMER PROTECTION & FINANCE
1982
GPO ITEM 1019-A, 1019-B (MICROFICHE)

TELECOMM EVAL ECON US
LEGAL

1097 TELECOMM POLICIES IN SEVENTEEN COUNTRIES: PROSPECTS FOR FUTURE COMPETITIVE ACCESS
1983
GPO ITEM 126-D-5 (MICROFICHE)

TELECOMM POLICY INTERNAT EVAL

1098 TELECOMMUNICATIONS FOR DEVELOPMENT
(WORLD COMMUNICATION YEAR)
1983
INTERNATIONAL TELECOMMUNICATION UNION
GENEVA, SWITZERLAND

TELECOMM ECON SOC ITU
RURAL

1099 TELECOMMUNICAATION DEVELOPMENT IN ASIA
1982
TRANSPORT AND COMMUNICATIONS BUL ASIA AND PAC
N55 P68-78

ASIA TECH TELEPHONE STAT

1100 TELECOMMUNICATION (DEVELOPING COUNTRIES)
JAN/FEB 1982
COURIER
P38-67

TELECOMM ECON

1101 TELECOMMUNICATION AND THE FINANCIAL SERVICES INDUSTRY
1984, OCT 1
TELEPHONY
207 N15

TELECOMM FIN BUSI

1102 TELECOMMUNICATION DEVELOPMENT IN AFRICA
1982
TELECOMMUNICATION POLICY
49 N10 P668-686

TELECOMM AF

1103 TELECOMMUNICATION IN AFRICA: A SURVEY
1977 JAN/FEB
CABLE LINES
5:/2

TELECOMM AF PANAFTEL CABLE
TECH RADIO ECON TECHNICAL

1104 TELECOMMUNICATION: UNE ERE NOUVELLE POUR L'AFRIQUE
1978 JN
ENTENTE AFINE
31:18

AF TELECOMM SOC

1105 TELECOMMUNICATIONS & INDUSTRIAL DEVELOPMENT: A STUDY OF PUBLIC PURCHASE, TECHNOLOGY TRANSFER IN IRISH TELECOMMUNICATION IN 1980'S
1980
NTIS PB84-117191

TELECOMM BUSI ECON TECH
IRELAND HIST

1106 TELECOMMUNICATIONS DEVELOPMENT PLAN FOR THE 80'S IN THE ROC
1981, APRIL
INDUSTRY OF FREE CHINA, TAIPEI
55 N4 P11

TAIWAN TELECOMM PLAN

1107 TELECOMMUNICATIONS IN LATIN AMERICA: STILL TO BE BUILT
1977, SEPTEMBER
IDB NEWS
P1-3
INTER-AMERICAN BANK: WASHINGTON, D.C.

L AMER TELECOMM

1108 TELECOMMUNICATIONS IN THE ARGENTINE REPUBLIC
1983
TELECOMMUNICATION JOURNAL
50 N12 P658-701

ARGENTINA TELECOMM

1109 TELECOMMUNICATIONS POLICY IN THE US AND
EUROPE: REGULATORY, SOCIAL AND POLITICAL ISSUES II
1983, SEPT
INFORMATION TECHNOLOGY AND PEOPLE (GB)
3 N5 P122-36

US EUR TELECOMM SOC
POLIT REG

1110 TELECOMMUNICATIONS: ECONOMIC STUDIES
CITATIONS FROM THE NTIS DATA BASE
1984, 83, 81
PB83-801670 & PB84-876317 & PB81-808883

TELECOMM ECON RES

1111 TELECOMMUNICATIONS: THE GLOBAL BATTLE
1983, OCT 24
BUSINESS WEEK
P126-131

TELECOMM WORLD

1112 TELECONFERENCE SYSTEM TO HELP HIGHER EDUCATION IN INDONESIA
THANKS TO RURAL SATELLITE PROGRAM
1984, MAY 14
TELEPHONY
206 N20 P152-8

TELECONF EDUC INDONESIA RURAL
PLAN

1113 TELEPHONE: A MASSIVE THIRD WORLD MARKET
1985
THIRD WORLD DEVELOPMENT
2 P18-24

TELEPHONE

1114 THE DEFT ANSWER TO RURAL TELEPHONY
1982, DECEMBER
TELEPHONY

TELEPHONE ECON RURAL

1115 THE ECONOMICS OF NEW EDUCATIONAL MEDIA VOL 2: COST AND EFFECTIVENESS
1980
UNESCO: PARIS

TECH EDUC ECON EVAL
COST

1116 THE ROLE OF TELECOMMUNICATION IN RURAL DEVELOPMENT
1983, DEC
TELECOMM (INDIA)
33 N2 P36-47

TELECOMM RURAL INDIA THEORY

1117 THE SITE EXPERIENCE. REPORTS AND PAPERS ON MASS COMMUNICATION
1983
MONOGRAPH
UNESCO: PARIS

SAT TV RES EDUC
UNESCO TRAIN SOC HIST

1118 THE SUDAN RURAL TELEVISION EXPERIMENT
1987, SEPTEMBER
EDUCATIONAL BROADCASTING INTERNATIONAL
11 N3 P113-116

SUDAN RURAL TV

1119 TOMORROW BELONGS TO THE U.S.: TELECOMMUNICATIONS TECHNOLOGY IS THE
WAVE OF THE FUTURE
1983, FEB 15
VITAL SPEECHES
49 P280-284

TELECOMM TECH US

1120 U.A.E. TELECOMMUNICATIONS
IN POLISHUK & O'BRYANT, EDS, TELECOMM & ECON DEVELOPMENT
1977
HORIZON HOUSE INTL: DEDHAM, MA

TELECOMM M EAST

1121 UNIVERSAL TELEPHONE SERVICE PRESERVATION ACT OF 1983:JOINT
HEARINGS S.1660 & OTHERS
1983, JULY 28-9

WORLD TELEPHONE DEREG POLICY

1122 WARC OF 1979: FINAL ACTS
1979
INTERNATIONAL TELECOMMUNICATION UNION: GENEVA, SWITZERLAND

WARC TELECOMM POLICY

1123 WORLDWIDE REPORT TELECOMMUNICATION POLICY, RESEARCH & DEVELOPMENT
1985 JAN
PB85-941604/XAB

TELECOMM POLICY RES WORLD

1124 WORLDWIDE REPORT TELECOMMUNICATION POLICY, RESEARCH & DEVELOPMENT
1979
GPO ITEM 1067-L-10 (MICROFICHE)

WORLD TELECOMM POLICY RES

Index

Key words used to classify the bibliographical entries are listed below. Abbreviations used in the text are shown in parentheses. Numbers refer to the entry number.

Geographical Regions

Technologies and Media

Content and Applications

RURAL cont'd

SOCIAL (SOC)

SPECTRUM

STATISTICS (STAT)

TECHNICAL

Organizations

The Artech House Telecommunication Library

Preparing and Delivering Effective Technical Presentations by David L. Adamy

ISDN — Concepts, Methods, and Systems by Peter Bocker

The Executive Guide to Video Teleconferencing by Ronald J. Bohm and Lee B. Templeton

The Telecommunications Deregulation Sourcebook, Stuart N. Brotman, ed.

Mobile Telephony by George Calhoun

Measurements in Optical Fibers and Devices: Theory and Experiments by Giovanni Cancellieri and Umberto Ravaioli

E-Mail by Stephen A. Caswell

The ITU in a Changing World by George A. Codding, Jr. and Anthony M. Rutkowski

Cryptology Yesterday, Today, and Tomorrow, Cipher A. Deavours et al., eds.

Machine Cryptography and Modern Cryptanalysis by Cipher A. Deavours and Louis Kruh

Design and Prospects for the ISDN by G. DICENET

Communications by Manus Egan

Introduction to Satellite Communication by Bruce R. Elbert

Television Programming across National Boundaries: The EBU and OIRT Experience by Ernest Eugster

The Competition for Markets in International Telecommunications by Ronald S. Eward

Telecommunication Transmission Measurements and Quality by John Fennick

Telecommunication Systems by Pierre-Gerard Fontolliet

Optical Fiber Transmission Systems by Siegfried Geckeler

The Cable Television Technical Handbook by Bobby Harrell

Traffic Flow in Switching Systems by Gerard Hebuterne

A Bibliography of Telecommunications and Socio-Economic Development by Heather E. Hudson

New Directions in Satellite Communications: Challenges for North and South, Heather E. Hudson, ed.

Communication Satellites in the Geostationary Orbit by Donald M. Jansky and Michel C. Jeruchim

World Atlas of Satellites, Donald M. Jansky, ed.

Design and Prospects for the ISDN by G. DICENET

Communications by Manus Egan

Introduction to Satellite Communication by Bruce R. Elbert

Television Programming across National Boundaries: The EBU and OIRT Experience by Ernest Eugster

The Competition for Markets in International Telecommunications by Ronald S. Eward

Telecommunication Transmission Measurements and Quality by John Fennick

Telecommunication Systems by Pierre-Gerard Fontolliet

Optical Fiber Transmission Systems by Siegfried Geckeler

The Cable Television Technical Handbook by Bobby Harrell

Traffic Flow in Switching Systems by Gerard Hebuterne

A Bibliography of Telecommunications and Socio-Economic Development by Heather E. Hudson

New Directions in Satellite Communications: Challenges for North and South, Heather E. Hudson, ed.

Communication Satellites in the Geostationary Orbit by Donald M. Jansky and Michel C. Jeruchim

World Atlas of Satellites, Donald M. Jansky, ed.

Handbook of Satellite Telecommunications and Broadcasting, L. Ya. Kantor, ed.

World-Traded Services: The Challenge for the Eighties by Raymond J. Krommenacker

Telecommunications: An Interdisciplinary Text, Leonard Lewin, ed.

Telecommunications in the U.S.: Trends and Policies, Leonard Lewin, ed.

Introduction to Telecommunication Electronics by A.Michael Noll

Introduction to Telephones and Telephone Systems by A. Michael Noll

Teleconferencing Technology and Applications by Christine H. Olgren and Lorne A. Parker

The ISDN Workshop: INTUG Proceedings, G. Russell Pipe, ed.

Integrated Services Digital Networks by Anthony M. Rutkowski

Fiber Optics Communications, Henry F. Taylor, ed.

Twisted Pair Network Technology by Randall W. Terpstra

Optimization of Digital Transmission Systems by K. Tröndle and G. Söder

Writing and Managing Winning Technical Proposals by Timothy Whalen

The Law and Regulation of International Space Communication by Harold M. White, Jr. and Rita Lauria

www.ingramcontent.com/pod-product-compliance
Lightning Source LLC
LaVergne TN
LVHW080041210926
848481LV00011B/3248